GIFTS FROM A GLACIER

The Quest for an American Flag

and 52 Souls

By

TONJA ANDERSON-DELL

Published by Richter Publishing LLC www.richterpublishing.com

Editors: Kati Scanlon, Katharina Jung & Monica San Nicolas

Book Cover Design: Jessie Alarcon

Copyright © 2017 TONJA ANDERSON-DELL

All rights reserved. No part of this book may be reproduced in any form by any electronic or mechanical means (including photocopying, recording or information storage and retrieval) without permission in writing from the author or publisher.

ISBN: 1945812133

ISBN-13: 9781945812132

DISCLAIMER

This book is designed to provide information on the history of a plane crash in 1952. This information is provided and sold with the knowledge that the publisher and author do not offer any legal or medical advice. In the case of a need for any such expertise consult with the appropriate professional. This book does not contain all information available on the subject. This book has not been created to be specific to any individual people or organization's situation or needs. Reasonable efforts have been made to make this book as accurate as possible. However, there may be typographical and or content errors. Therefore, this book should serve only as a general guide. This book contains information that might be dated or erroneous and is intended only to educate and entertain. The author and publisher shall have no liability or responsibility to any person or entity regarding any loss or damage incurred, or alleged to have incurred, directly or indirectly, by the information contained in this book or as a result of anyone acting or failing to act upon the information in this book. You hereby agree never to sue and to hold the author and publisher harmless from any and all claims arising out of the information contained in this book. You hereby agree to be bound by this disclaimer, covenant not to sue and release. You may return this book within the guarantee time period for a full refund. In the interest of full disclosure, this book contains affiliate links that might pay the author or publisher a commission upon any purchase from the company. While the author and publisher take no responsibility for any virus or technical issues that could be caused by such links, the business practices of these companies and or the performance of any product or service, the author or publisher has used the product or service and makes a recommendation in good faith based on that experience. The author has tried to recreate events, locales and conversations from their memories of them. In order to maintain people's anonymity in some instances the names of individuals and places may have been altered. Some identifying characteristics and details such as physical properties, occupations and places of residence may have been changed. The opinions and stories in this book are the views of the author and not that of the publisher.

Table of Contents

ACKNOWLEDGEMENT	5
DEDICATION	6
INTRODUCTION	7
Chapter 1 - The Beginning...December 1999	10
Chapter 2 - The Accident Report	17
Chapter 3 - Dying	27
Chapter 4 - Second Chance	35
Chapter 5 - Making it Through the Fire	39
Chapter 6 - The Plane Has Been Found	64
Chapter 7 - The Battle	72
Chapter 8 - Coming Home	86
SYNOPSYS	111
SOLDIERS STILL MISSING	125
ABOUT THE AUTHOR	126

ACKNOWLEDGEMENT

This book is an expression of the steps I have taken to get a flag for my grandmother, the experiences and the journey to find and bring home the Soldiers aboard the Missing C-124 Globemaster from November 22, 1952. This journey was a long one and could not have happened without the support and guidance of my friends, family members of the soldiers, and my family. Along the way, I have gathered the treasures each of you provided and contributed. I have been enriched by each of you in different ways. I would like to acknowledge all the family members of the fallen 52 soldiers on the Globemaster plane who kept pushing me to put all my research into a book for the world to read. For those that allowed me to attend the services of their fallen soldier. To Alaska Army National Guard Pilot Captain Brian Keese and his team. Without you all taking the time to turn around research what you saw, the crash site would still be unfound. For this, I am forever grateful. To all of the men and women from Joint POW/MIA Accounting Command (JPAC), Joint Base Elmendorf–Richardson (JBER), Armed Forces Medical Examiner (AFME) and Allen Cronin with Air Force Mortuary Affairs Operation (AFMAO) who put their lives on the line to bring home our fallen soldiers, Thank you for all your hard work. To Sean Doogan with Alaska Dispatch News for helping me get answers and keeping our story alive. To my best friend, Janet Strohsack, she's been my road dog. She has been a positive reinforcement for me during this long journey. For this, I say thank you. Special thanks to my mother Grace Dupree, my father Isaac Anderson Jr. for pushing to fight for answers. My brother Ernest Dupree, my children Destinee (Sean) Anthony, Tevin Valdez, Faith Lark, Angel Dell and Angela Dell. My Grandchildren Chase and Eli. Children, you all were my cheerleaders and told me to never give up. I hope I have made you proud. To my husband Earnest Dell, you allowed me to fulfill my dream. Through all the long nights of research, letter writing, phone call, and traveling you gave me strength, support, and word of comfort. I could never be able to put into words how blessed I am for having you by my side.

DEDICATION

    I would like to dedicate this book to my grandparents Dorothy M. Anderson and Airman Isaac W. Anderson Sr.

    Grandma, I started out in search of a flag for you, and I found so much more. I wish you had lived long enough to see I received a flag for you but I know you have seen it all through heaven's floors.

    To my grandfather I say this "To never know a person but so driven too; is a powerful thing. Sir, I am so proud to be your granddaughter and I hope I have made you proud too."

    To the men aboard the missing C-124 Globemaster, you all made the ultimate sacrifice, and I will fight until ALL 52 of you have returned home to your families.

INTRODUCTION

This is the story of the 52 United States soldiers on a C-124 Globemaster plane that went missing on November 22, 1952. These soldiers were returning from leave, missions, or arriving at their new base.

The US Military prides itself with their dedication to never leaving their fallen soldiers behind. This was not the case of this missing plane and the 52 souls, who dedicated their life to the US. The men on board were left frozen where they had fallen.

It has been stated by the military that anything was feasible if someone was willing to spend the time, money, and energy on it. Even locating and rescuing these men. The US Military decided that a search mission was too dangerous and not worth the time, energy and money to bring the fallen soldiers home.

In 1999, I turned to a widow of one of the men on that flight—my grandmother. I asked if I could search for the truth of her husband and the other soldiers' death. Until this point, she had not finalized my grandfather's affairs. She replied to me "tell them I am ready for my American flag now." The flag that honors those who have served in the military. She had to wait over 50 years to receive that honor.

I never knew who my grandfather was. He joined the military to provide for his family and died when the C-124 Globemaster he was on crashed into a mountain. I found it very noble of him to join the military during a time when being an African American male was hard and providing for a family was even harder. The military offered good benefits, and a means to provide for his family.

Airman Anderson and 51 other soldiers boarded a C-124 Globemaster that was headed to Anchorage Alaska when it lost contact and crashed into a mountain peak. These soldiers joined the military for a better life and to serve their country. However, when it came time to search for the missing soldiers, the Air Force asked if it was feasible to recover objects there. This indifferent attitude the Air Force showed my grandfather , and the other 51 soldiers, when it came to searching for that missing plane is what inspired me to start this journey of finding closure.

Air Mobility Command MUSEUM

**The AMC Museum's C-124A, not the actual 51-0107 that was lost.
This is the only C-124A plane left.**
Photo Courtesy of the AMC Museum.

Chapter 1 - The Beginning...December 1999

When I asked my father, "Could you tell me anything about granddad?" He sighed deeply and replied, "Tonja; I don't know anything. I was always told he died while in the military. His plane crashed, I think." This was the family story, but no one knew if it was true or not. "Go ask your grandma," he said.

Over the next couple days, I tried to find a way to bring this up to my grandma without reminding her of the hard times, difficult memories, or the pain that she likely still carried deep down in her heart. So, I planned to meet with her over the weekend.

I arrived at my grandmother's house feeling very nervous to ask her about granddad, but how else would I get the answers? "Grandma, can I talk to you? I have a lot of questions about granddad."

Her face showed signs of concern, "Why, Tonja? You're always looking for something to get into!" I was the one grandchild who was always trying to find out our ancestors' history and where we came from. I guess for me to know who I am, I'd need to know where I came from. She sat with me and told me everything that she could remember about the day she'd lost the love of her life and became a widow.

"Tonja baby," she said, "I was a young wife and a young mother. Your grandfather joined the Air Force to provide for us during a time when blacks struggled to have the same civil right as whites; such as equal employment, housing and education. He felt that the military was the best place to provide for us during these times." She paused for a moment and then continued, "I remember this like it was yesterday. He was home from leave for only about a month. We talked and didn't have a good feeling about him returning to his duties with the Air Force. He spoke about what he witnessed, being cold, and all the planes that had crashed. At that point, we decided he would not go back to the base. I was going to try everything to stop the Air Force from sending him back to Alaska. I filed paperwork and did whatever I could to stop the transfer; because of this, he was considered AWOL (a military term meaning Absent Without Leave) as he did not return back on time. Going AWOL is a serious offense, and you can be arrested and serve time in jail. We did not want this to happen, so he had to return to duty to McCord Air Force Base on October 22, 1952. I waited at home to hear how bad things were for him after not reporting back in time. I didn't hear from him for a long time, and I started to worry. I was in the kitchen listing to the radio when the news broke. 'Military plane has gone missing in Alaska.' I knew it was his plane the moment I heard it. I just knew it!" I looked at her as she said, "And I thought, God, please tell me I'm wrong.

I eagerly interrupted her—"Grandma," I inquired, "Did they ever find the plane? Did he ever come home?" She looked at me and touted, "Shut up, let me finish." I laughed, and she continued.

I did not hear anything from the military for a while. I sat by the radio and listened for any sign of hope. Our family and my friends came together for prayer, and they became my support system during this time. I remembered a couple of days had passed and I was thinking of your grandfather and crying because we didn't want him to go back and now his plane was missing; your father's second birthday was the next month, and the tears were pouring. I pulled it together to get the mail to see if I'd gotten an update from the military, instead I got the surprise of my life. It was a letter from your grandfather; he'd put it in the mail hours before he'd boarded the plane headed to Alaska.

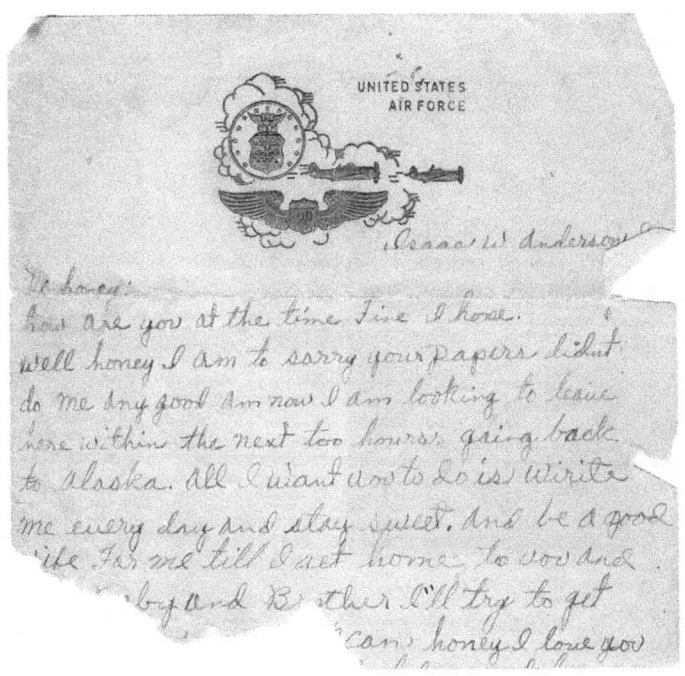

The letter Airman Isaac Anderson placed in the mail to his wife.

The next time the military communicated with me was in a letter dated December 20, 1952, almost a month after the crash. This letter served as my notification from the Air Force, telling me my husband was killed in the plane crash and how sorry they were for my loss." My husband died one month after he returned from leave; one month to the day, Tonja.

I then asked her, "Do you mind if I look into what happened to Grandpa and the plane crash? I don't want to bring up any feelings or upset you." She replied, "Tonja, I've lived all my life thinking that since they never found the plane and or any bodies, he'd still walk back through the door one day. It's now been forty-something years, so I guess he's never coming through the door. I give you permission to look into finding some answers on what happened to the plane and everyone on it. I just ask that in your search and investigation you let the USAF know I'm ready now...I'm ready for the flag."

"Grandma," I probed, "What do you mean?" She then explained, "Tonja, back then, the flag was something they gave to the families when one of their loved ones died while serving their country. He's not coming back, and I'm now ready for that flag."

```
                625TH AIRCRAFT CONTROL SQUADRON
             APO 942 c/o Postmaster, Seattle, Washington

                                                     20 December 1952
```

My Dear Mrs Anderson,

It will be difficult to convey to you the feeling prevailing in this organization upon learning that Isaac W Anderson Sr had been killed in an aircraft accident which was on a flight from Mc Chord Air Force Base, Washington to this base. As deep as that feeling is here, we know it can never be exceeded by that at home. With this thought in mind, I and the members of this organization convey to you our most sincere condolence.

Isaac was returning from a morale leave when the accident occurred. The plane he was on, crashed between fifty and sixty miles from Elmendorf Air Force Base, Alaska, its destination. All the facts concerning the incident and the ill fated plane are not known and some can never be learned. A civilian doctor flew to the scene of the accident 28 November 1952 and surveyed the disaster area, which is in a very mountainous region. There were no survivors. The casualty which cost us the life of a fine Airman, occurred in an instant and no physical or mental anguish was experienced. A ground rescue team has started to the scene of the accident and upon their return more information of the accident may be known.

Losses such as these can never be adjusted and that we all painfully know. The pain of loss can be alleviated somewhat by the knowledge that before death, life had been in keeping with the ideals of the Armed Forces and the nation which we serve.

Isaac's personal effects will be safeguarded and prepared for shipment as soon as possible.

Please feel free at all times to communicate with me or any member of this organization and be assured our sentiments run deeper than can ever be stated.

                              Very sincerely yours,

                              ELLSWORTH VALENTINE
                              Captain   USAF
                              Commanding

"Grandma, I don't know when or how, but I promise I will not stop until I find out what happened to Granddad and the plane. I will get you your flag."

Following December of 1999, I spent the next several months researching, writing letters, emails, and calling anyone and everyone who would pick up their phones. I had no clue what I was doing or where to even start, but I did know I was not giving up until I got an answer. I first wrote to the USAF asking for any information on the plane crash and a copy of my grandfather's military records.

To whom this may concern,

My name is Tonja Anderson. I'm looking for information on the death of my grandfather. His name is Isaac W. Anderson Sr. He died on Nov 22, 1952 in Alaska from an aircraft accident during the Korean Conflict. In MacDill Air Force Base, there was a hanger that had his picture along with the other crewmen that died in it. I just want know more about this accident and if the Air Force ever recover his body or the plane. My grandmother cannot tell me that much about what happened.

Thanks,
Tonja Anderson (granddaughter of Airman Anderson)"

Air Force Association

Dear Sir,
My grandfather Isaac W. Anderson Sr. of the USAF, who was member of your association, has been missing on Mount Gannett since November 22, 1952.

He and 51 other men were in flight from McChord Air Force Base to Elmendorf Air Force Base. I'm trying to contact the families of the other men aboard this plane. I wonder if you could help me find out last known name and address of their next-of-kin. The list of men who were aboard the plane is:

Col. Eugene Smith,A/2C Thomas C.Thigen,Capt.Walter P. Tribble,Capt. Robert W. Turnbell,T/Sgt. Leonard G. Under,S/Sgt. R.D. Van Fossen,A/2C. B.F. White, Lt. Col. S. Singleton,2nd Lt. Reginald Bule,Pvt. Leonard A. Kittle, Pvt. James green, 2nd Lt. Alan Berger, 2nd Lt. Edwin H. Loeffler,M/Sgt. Edward J. Schnore, Maj. Earl J. Sterns, Basic Airman Isaac W. Anderson Sr.,A/2C Verne V. Bubahn,Capt. William N. Coombs,A/2C Thomas J. Condon,Capt. Delbert D. Draskey,A/2C Carroll R.Dyer,Capt. Jerome H. Goebel,A/1C Marion E. Houton,Lt. Jack R. Leaford jr.,A/2C Dan F.McMann, A/3C Lloyd L.Mattews,A/2C Edmon W.Mize Jr.,Col.Noel E. Hoblit,2nd Lt. Robert E.Moon,A/1C Sterling E. Newsome,S/Sgt James H.Ray Jr.,1st Lt. Donald A.Sheds,and Capt. John E.Ponikvar.

If you could find out anything for me, I would greatly appreciate it. I have heard that your association is very successful in finding this information and so I decided to appeal to you to help me locate the families of these men, to ask them to help get the Air Force to go back and excavate the plane and its remains.

I will appreciate any attention you show me.
Thank you,

Mortuary Affairs Division
San Antonio, TX

Dear Ms. Anderson,

Your email of 9 September 2000 was forwarded to this office for response. Due to the Privacy Act, we cannot provide names, addresses, and telephone numbers of family member of those killed in the C-124 mishap. There is currently no plan to begin recovery operations. I'm sorry that I cannot be more helpful to you in this matter.

-Mortuary Affairs Division

I also searched the internet and recruited people to help me look for information on any living Isaac W. Anderson Sr. I searched for him just in case the military records came back with a small piece of hope that he would still be alive.

```
Subj:   Re: Information
Date:   8/14/00 12:47:52 PM Pacific Daylight Time
  .n:   ORANGEE123
To:     TPubzmxdd

Tonja, I found 2 entries in the AOL White Pages for that same name: Isaac W Anderson. They were listed as living, so don't
get your hopes up. But if I were you I would want to call them. These white pages ignore the Sr. or Jr. on a name.

       N Cambridge Rd
Saucier, MS 39574


       Warren St
Shelby, NC 28150

Please let me know what you find out.
```

After reading, the email I felt a little overwhelmed. Could he really be alive? Did something happen during the crash that caused him to lose his memory? I mean the military never found the crash site until days later, and they never found/identified remains of the men.

I went to the Social Security website, tried to see what I could find on him and if anyone has filed for his benefits other than my grandma. That turned out to be a dead end. So I decided to call the men listed in the white pages with the same name as my grandfather. I sat down in my living room, gathered my nerves, and picked up the phone. My fingers shook as I punched the numbers into the keypad. The ringing of the phone was loud in my ear, and my heart started racing, scared that I might actually hear my grandfather's voice on the other end.

Me: Hello, may I speak with Isaac W. Anderson?
Person answering the phone: I'm sorry, he's not home. Can I help you?
Me: I'm looking for Isaac W. Anderson, who might have served in the USAF. Did your Isaac W. Anderson serve in the USAF?
Person answering the phone: I'm sorry, but he did not. I wish I could help you.
Me: Thank you very much, goodbye.

Just making that first short phone call took all my energy. I was so scared that he might pick up the phone and say, "Yes I am, how I can help you?" At that point, what would I say next?

I waited until the next day to make the second call.

Ring...ring...ring...

Person answering the phone: Hello
Me: Hello, is Isaac W. Anderson home?
Person answering the phone: I'm sorry, but he's no longer at this number.
Me: Thank you and have a nice day.

Don't worry...the small piece of hope didn't last long before I realized that it was a hope of grandma's and a dead end for me. For a moment, I, too, had some hope of finding him alive for her.

I sat down and read through all of the replies I'd received from my emails about my grandfather. The one I received from a Ms. Lynn from the Air Force Historical Research Agency at Maxwell Airforce Base, was very interesting.

June 29, 2000
Ms. Anderson;
"Thank you for your e-mail. I did locate the aircraft accident report. It is only in microfilm and rather difficult to read. I was unable to print a readable copy. However, if you will send me your mailing address I will send you a copy of microfilm which you may read at any local public or college library.
Sincerely
Ms. Lynn"

Excited, I wrote back to Ms. Lynn.

Ms. Lynn,
Thank you very much, my address is listed below. I look forward to seeing the report. Thank you for taking time out to help me.
Thanks
Tonja Anderson

On July 6, 2000, I received a package in the mail from Ms. Lynn. Inside the package was an unreadable printed copy the accident report and the full accident report on microfilm with a little note explaining what was inside. I know she might have thought she was just doing her job but for me it as so much more. I was like a kid in the candy store, jumping for joy.

Chapter 2 - The Accident Report

I sat down and tried to read the printed copy of the report, but it was very hard because it was so old. I decided to go down to my local library to see the report on microfilm. At the library, I viewed the first several pages only to be blown away by what I read. I printed all 300 pages of the report, went home, and read page after page. I stayed focused on several different things: the passenger manifest (I needed to see if Anderson was on the list), the radio communication report, the statement from the forecasters, and the statement given to Maj. Potter, Mr. Kieffer, Lt. Sullivan, and Dr. Moore.

The first items to catch my eye were the statements from Lt. Sullivan and Dr. Moore. The crash took place on November 22, 1952, and Lt. Sullivan and Dr. Moore were the individuals assigned by USAF with the duty of locating and identifying the possible crash site. Lt. Sullivan and Dr. Moore located the crash site on November 28th, some six days after the loss of communication with the men on the plane. Lt. Sullivan stated that he left Elmendorf Air Force Base with Dr. Moore flying the plane. They were headed to Serpentine Glacier as it was first believed the plane had crashed there. This was based on a map provided by the 10th Air Rescue. They would later search Surprise Glacier on the slopes of Mount Gannett and see the tail section of an aircraft on the glacier floor.

Upon landing on the soft snow that covered a large part of the glacier, they went directly to the tail end of the plane. The snow around the wreckage was dry and about six to eight feet deep over the floor of the glacier, drifting in many areas to a greater depth. They would make it close enough to the tail to visually see the numbers 1107 on the right side of the vertical stabilizer of a C-124-type plane. With those numbers, they were able to positively identify this piece as part of the C-124 Globemaster that went missing on November 22, 1952. The tail section of the plane had been seared off from the fuselage and was tilted forward on the glacier. Dr. Moore saw the tail and stabilizer, commented, "The impact G's must have been tremendous." They would see a blanket caught on the left elevator of the tail section. They looked for human remains near the tail end but failed to locate any. They were standing near the tail section and looked up the mountain to see mounds of snow with metal protruding through. They decided to climb 150 feet above the tail and noticed a blanket partly covered in frozen blood. The sunlight had started to melt the snow and caused a stench of decomposing flesh to fill the area. Dr. Moore proceeded back to the plane while Sullivan continued to search around. Sullivan stated that he saw a dark section of snow and went to it, digging about two to three feet before he pulled up a military parka that had no identification. Strangely, it was still buttoned up yet there was no human remains to be found in or around it. The front of the parka was charred through one layer of cloth. They stated that the plane and its contents were spread across two acres and under snow, at least eight feet deep in certain areas

The tail end of the C-124 Globemaster

Lt. Sullivan standing next to the tail end of the C-124 Globemaster

Dr. Moore's statement was very interesting. He gave an account of what he saw while on the glacier with Mr. Sullivan. He was asked if it was feasible to recover objects there and his response recorded in the report stated, "My answer would be that anything is feasible if one is willing to spend enough energy, time and money on it, ranging from the thousand to a million to ten million to a hundred million to a billion dollars, anything can be done within reason. Thus, certainly, everything could be recovered if one wished to. It is a question, in my opinion, whether the objectives are worth x number of dollars and energy. It looks to me as if it were a job for about a month's time, for eight or ten men, being provided with food by aerial supply, working full time on the job to reasonably excavate the remains of that wreck. I conferred at some length with Lt. Hackett of 10th Search and Rescue, regarding this. He's an experienced mountaineer in my opinion, and I would feel that his opinion regarding this matter is just as good, if not better than my own as to procedures and making a recovery by the ground party."

As I read this; tears rolled down my eyes and fire filled my soul. How could someone put a price tag on another human being's life? I felt my grandfather's life and the lives of the other men on the plane were being measured based on financial cost. These men gave everything for their country, and this man stated: "…anything is feasible if one is willing to spend enough energy, time, and money on it." My grandfather and the other 51 men's lives were worth more than any dollar amount the government would or could put on their lives. Our government can spend money on unjust causes, but they were willing to put a price on bringing home their fallen. I spent the next couple of weeks going through a range of emotions. How could the land of the free and the home of the brave not care about our men? I was brought up believing that the government would always do what's right for our soldiers. That the great USA is a country built on the foundation of integrity. If you enlist and give your life over to serve, that you will be taken care of. That you will be honored. Now I just felt as though I had been lied to. I was more determined than ever to find my grandfather's remains and make the government do what was right. Bring the men home for proper funeral services

I sat down and drafted several letters that day, to my local Tampa, Florida Senators and Congressman; Senator Connie Mack, Senator Bob Graham, and Congressman Jim Davis.

Senator Connie Mack,
1342 Colonial Blvd.
Suite 27
Ft. Myers FL. 33907

Senator Mack;

I am writing you about the death of my grandfather Isaac W. Anderson. He died during the Korean Conflict. The aircraft that he was on crashed into Mount Gannett in Alaska. The Air Force never found his body nor the other 51 men on board. I have wrote many different people trying to get information about that accident. The only reply I received was from a Mrs. Lynn Gamma. She sent me a copy of the accident report. In a statement to the review board a Dr. Moore and Sullivan suggested how to excavate the aircraft and it's remains. He make this statement again in the Fairbanks News Paper on Dec. 01,1952. They sent me a copy of the paper. I want to know how I can find out if the Air Force ever went back to excavate the plane and it's remains. I though that it was said that "no one will be left behind and that the United States will always try to get you back". If this is true, why didn't they go back and get 52 of their men who has fallen?

I have been watching the news and seen the crashing of the Concord. I seen where they are going to remove the remains of the people in this plane. I do understand that the weather is bad in Alaska, but that would make it a little better. The snow would help keep the remains iced.

In a conference with Major Dwight H. Potter and Mr. Wm.L.Kieffer a question was asked "If it is Feasible to recover objects there"? The answer was" anything is Feasible if one is willing to spend enough energy, time, and money on it. Ranging from 10,000 to 10 million dollars, anything can be done within reasons. Thus certainly everything could be recovered if one wish to. It I a question, in my opinion, whether the objects are worth x number of dollars and energy. I think that my grandfather as well as the other 51 people on this aircraft are worth more than x number of dollars. How can the Air Force put a price on someone's life /remains? What I am looking for is a closer in my life about the remains of my grandfather. Did the United States Air Force go back and get their aircraft and it's remains?

My father is going to be 50 yr. old on Dec.25th. The gift I want to give him is how his father died and if the Air Force ever went back to get his father's remains. My father was 1 ½ yr. old when his father died. He doesn't even know what kind of father he had or if his father dying was worth something. I lost a grandfather that year also. I will never know the man he was or could have been and

to me this means more than any dollar the Air Force could have spent trying to bring his remains home to his family.

Thank You

Tonja Anderson

DEPARTMENT OF THE AIR FORCE
WASHINGTON, DC

OCT 2 0 2000

Office of the Secretary

October 12, 2000

The Honorable Jim Davis
United States Representative
3315 Henderson Boulevard, #100
Tampa FL 33609

Dear Mr. Davis

    This responds to your inquiry for Ms. Tonja Anderson regarding her inquiry concerning the recovery of her grandfather's body from an Air Force C-124 Globemaster crash in Alaska in 1952.

    Air Force Mortuary Affairs officials advise that, due to the inaccessible location of the wreckage and the terrain, there have been no attempts to recover the wreckage or the remains and there is currently no plan to begin any recovery operations. Should Airman Anderson's body eventually be recovered and identified from this crash site, a funeral with appropriate military honors will be conducted.

    We sent a similar letter to Senator Connie Mack and trust this information is helpful.

Sincerely

MICHAEL K. GIBSON, Lt Col, USAF
Deputy Chief, Congressional Inquiry Division
Office of Legislative Liaison

Office of The Deputy Assistant
Secretary of Defense for POW/MIA
2400 Defense Pentagon
Washington DC 2033-2400

Dear Sir.

I am writing this letter about the death of my grandfather, Airman Isaac W. Anderson. He died in a aircraft accident on Nov.22,1952 in Alaska on Mount Gannett. In the accident report the Air force held a conference. In this conference a Dr. Moore make a statement about what he saw and then he suggested to the board how to excavate the plane and it's remains. He makes this statement again to the Fairbanks News Paper. Then a Major Potter was asked "If the objects there feasible"? The answer was anything is feasible if one is willing to spend enough energy ,time and money on it,ranging from 10,000 to a million to ten million to a hundred million to a billion dollars, anything can be done within reason. Thus certainly everything could be recovered if one wished to.It is a question whether the objects there are worth x number of dollars and energy ."

What I want to know is how can the Air force allow someone to put a dollar sign on someone life/remains? I feel that my grandfather as well as the other 51 people aboard this plane is worth more than x number of dollar the Air force could have spent trying to bring the remains home to their families. Let me make it clear that I know the weather played a part in the search and recovery of this aircraft, but the Air force could have went back the summer of the next year. After reading this report I hold a lot of anger toward the military. These men went into the Air Force to protect and serve their Country and for this their Country turned their backs and thanked them by leaving the remains of their fallen men on a mountain side.

I wrote to my Congressman and my Senator for help in this matter. They have received a letter from a Michael K. Gibson, Lt. Col,USAF. In this letter their stated that "due to the inaccessible location of the wreckage and the terrain,there have been no attempts to recover the wreckage or the remains". I have the accident report and in it there are pictures of Dr. Moore next to the plane's tail. If this site was inaccessible how was these pictures taken and how did the Air force identify this as the missing C-124 that went down?

What I want is to know why haven't the United States Air Force gone back to excavate this plane and it's remains?

Sincerely

Tonja Anderson
The Grand Daughter of Airman Isaac W. Anderson Sr.

Letter after letter, I got the same dreadful reply:

"Air Force Affairs officials advise that, due to the inaccessible location of the wreckage and the terrain, there have been no attempts to recover the wreckage or the remains and there is currently no plan to begin any recovery operation."

I was not giving up until someone gave me the answer I was looking for.

"Ms. Anderson, the USAF has decided to put together a team to go back to Mount Gannett and see if anything has surfaced over the past 40 years."

On October 18, 2000, I got a letter in the mail from someone trying to get me to understand the severity of things during that time. I would later have the utmost respect for this man. Through all my letters and emails he never made me feel as if this was a fight I could not win.

DEPARTMENT OF THE AIR FORCE
PACIFIC AIR FORCES

18 October 2000

Mr. John H. Cloe
Office of History, 3rd Wing
10425 Kuter Ave Suite 320
Elmendorf AFB AK 99506-2630

Ms Tonja Anderson
7210 N. Manhattan Ave
#2121
Tampa FL 33614

Dear Ms Anderson

I apologize for the delay in getting back with you. I have been on a three week leave and returned to duty on the 16th.

Enclosed is the accident report I promised. Hopefully it will be of some help, although I note that you already have much of the information it contains. I found a map, which might be of some use to you and have made notations on it based on Lieutenant Sullivan's descriptions. As noted, Surprise Glacier is out of the area. He must have been referring to the unnamed glacier immediately to the south of Mt Gannett.

From what I could discern, the C-124 was off course when it flew into the west ridge of Mt Gannette. The planned flight path would have taken it over the community of Whittier to the south direct into Elmendorf AFB as shown in the enclosed second, smaller map.

I can only second guess the decision made by those in charge at the time not to search for the bodies of your grandfather and the others. They were probably considering the risk of further lives in their deliberations. The area around Gannett Mountain is heavily glaciated and dangerous. I feel that a search would have been made if the plane crash had occurred today. Our rescue forces are much better equipped and trained to deal with such tragedies.

We will continue to look into the matter and will stay in touch. Please let me know by e-mail when you receive the package and keep us informed on what you have found out.

I appreciate your concerns about your grandfather, and will do what I can to be of assistance.

Sincerely

JOHN H. CLOE
Chief of History

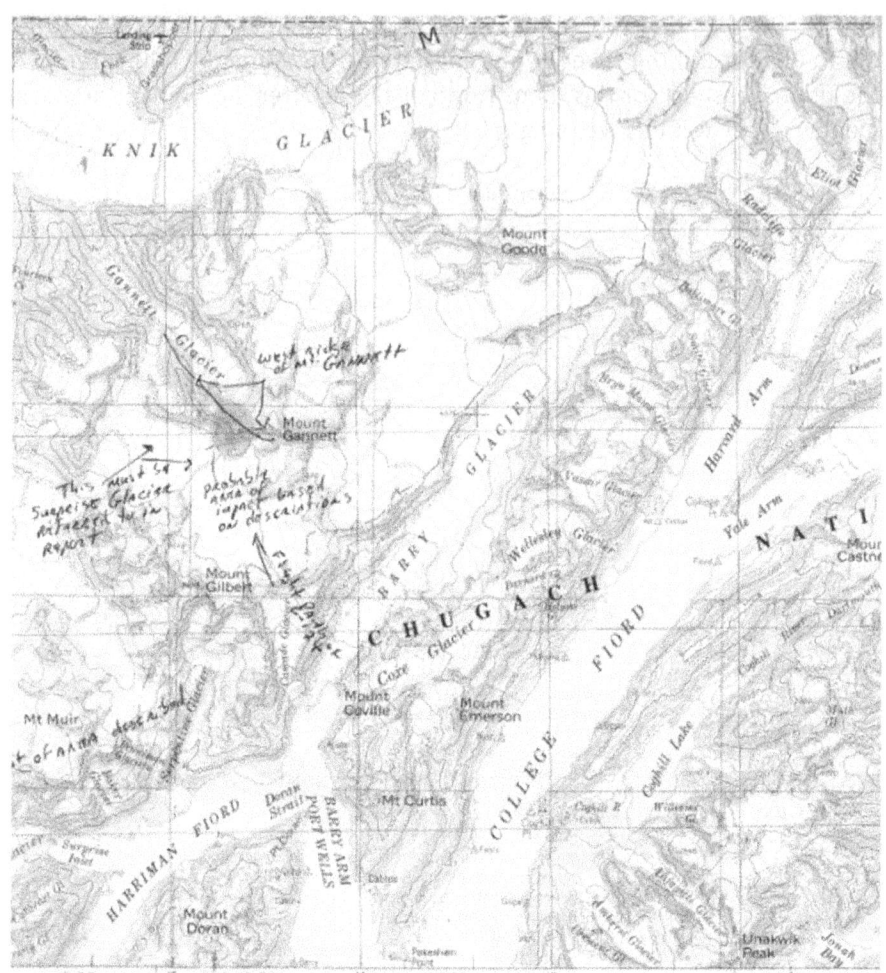

This copy of the map provided to me details the area and the plane crash

While I waited for answers to all the letters I had written, I wrote a request under the Freedom of Information Act (FOIA) to the United States Air Force for a copy of my grandfather's military records. I figured if I could not find out what happened to him in the crash that maybe his file would tell me who he was as a serviceman. A week or so after writing these letters I recieved a letter that would upset me for days afterward.

I had spent the last year trying to find out more on a man I have never met only to receive this news. About who he was before he died and what he was doing for the Air Force. Was he a good airman? Where do I go from here? How will I find out who Airman Anderson is now...? I just dropped my head, and I cried.

DEPARTMENT OF THE AIR FORCE
HEADQUARTERS UNITED STATES AIR FORCE
WASHINGTON D.C.

HQ USAF/ILV
1770 Air Force Pentagon
Washington, DC 20330-1770

0 2 NOV 2000

Ms. Tonja Anderson

Tampa, FL 33614

Dear Ms. Anderson

This is in reply to your DEFENSE LINK e-mail concerning the recovery of the wreckage from your grandfather's aircraft on Mt. Garrett in Alaska.

Mr. Gary Huey from the Air Force Mortuary Office has spoken with you recently. As he mentioned, there are currently no plans to conduct any recovery operations. Since this crash also included members from the Army and the Navy, our mortuary personnel have contacted those branches to determine if they have any plans for recovery. Unfortunately they do not.

We have also contacted the Central Identification Laboratory, Hawaii (CILHI) who is aware of this accident and would be responsible to initiate the recovery operation. You may want to discuss your concerns with them, by contacting them directly at:

CDR USA CILHI/TAPC-PED-H, ATTN: Dr. Tom Holland
Building 45, 310 Worcester Ave
Hickam AFB, HI 91853-5530
Telephone: (808) 448-8062 x 151

Should any new information become available, our mortuary staff will contact you.

Sincerely

MICHAEL J. KELLY, Colonel, USAF
Deputy Director of Services
DCS/Installations & Logistics

DEPARTMENT OF THE AIR FORCE
HEADQUARTERS AIR FORCE PERSONNEL CENTER
RANDOLPH AIR FORCE BASE TEXAS

0 7 FEB 2001

HQ AFPC/MSIMD
550 C Street West Ste 48
Randolph AFB TX 78150-4750

Ms. Tonja Anderson
7210 N. Manhattan Ave., Apt 2121
Tampa FL 33614

Dear Ms. Anderson

    This is in response to your Freedom of Information Act request (FOIA #01-0182) of 14 August 2000 requesting information from your deceased grandfather's records. Your request was forwarded to the Air Force Worldwide Locator for a reply to you.

    Upon your grandfather's untimely death in 1952 his records were forwarded to the National Personnel Records Center in St. Louis. Further research revealed that his records were burned in the fire at that facility in 1973. The address for the National Personnel Records Center is: NPRC/Air Force Reference Branch, 9700 Page Avenue, St. Louis Missouri 63132-5100 if you would care to contact them for any further assistance they might be able to render. The Air Force Locator is unable to assist you in obtaining the documents requested.

    Sincerely

    Betty-Leone C. Stewart
    BETTY-LEONE C. STEWART
    Management Assistant

    I had spent the last year trying to find out more on a man I have never met only to receive this news. About who he was before he died and what he was doing for the Air Force. Was he a good airman? Where do I go from here? How will I find out who Airman Anderson is now...? I just dropped my head, and I cried.

Chapter 3 - Dying

As I was losing hope of finding anything out about who my grandfather was, I started concentrating on why the plane crashed in the first place. As I looked over the passenger manifest of the soldiers on board that day, it gave me a renewed hope and I decided that I wanted to keep pushing forward to bring justice for all the lost souls.

Passenger Manifest:

Last Name	1st Initial	Rank	Crew Position	Last Name	1st Initial	Rank
Duvall	K	CAPT	AC (in comd of A/C)	White	B	A/2C
Cheney	A	CAPT	1st P	Cody	D	A/1C
Turner	W	1/LT	NAV	Martin	H	A/3C
Hagen	E	TSGT	IE	Ray	J	SSGT
Sprague	C	A/2C	2nd Eng	Thigpen	T	A/2C
Costley	E	SSGT	2nd Eng	Miller	E	A/2C
Owen	R	A/2C	Radio Op	Dyer	C	A/2C
Scott	M	A/3C	Radio Op	Budahn	V	A/2C
Ingram	G	A/1C	Loadmaster	Lyons	T	A/2C
Kimball	J	A/3C	Flight Attendant	Matthers	L	A/3C
Jackson	W	A/3C	Flight Attendant	Burns	B	A/2C
Smith	E	COL		Mize	E	A/2C
Singleton	L	LT COL		Hooton	M	A/1C
Stearnes	E	MAJ		Van Fossen	R	SSGT
Jackson	W	MAJ		Mc Cmann	D	A/2C
Tribble	W	CAPT		Newsome	S	A/1C
Ponikvar	J	CAPT		Condon	T	A/2C
Turnbull	R	CAPT		Kittle	L	PVT
Goebel	J	CAPT		Coombs	W	CAPT
Draskey	D	CAPT		Anderson	I	A/S
Sheda	D	1/LT		Card	R	PVT
Leadford	J	2/LT		Green	J	PVT
Moon	R	2/LT		Hoblit	N	COL
Berger		2/LT		Seeboth	A	CDR
Buie	R	2/LT		Schnore	E	MSGT
Loeffler	E	2/LT		Unger	L	TSGT

DETACHMENT 4, 4TH WEATHER SQUADRON
McChord Air Force Base
Tacoma, Washington

26 November 1952

STATEMENT

I came to work on the upper air charts at the MATS Weather Station at 1200P, 22 November 1952, and learned from Sergeant Holcomb, the other forecaster on duty, that three trips were going out: an RCAF North Star at 1400P to Elmendorf, a C-124 at 1500P to Elmendorf, and a C-54 to Kodiak at 1415P. Sergeant Holcomb had already made the flight level wind forecasts for these flights. Sergeant Holcomb was making up the flight folder for the North Star flight and when he finished he briefed me on the weather for that flight and then departed for lunch at approximately 1230P. The RCAF crew came in a short time later and was briefed by me. Later an Air Force pilot who did not identify himself except to say that he was going to Elmendorf came in and asked about the route weather. Using the maps and sequences I briefed him regarding weather along Military Airways and along Amber One Airway and also on typical wintertime weather conditions. In regards to the weather along Military Airways, I told the pilot that he would have instrument conditions and icing on the latter third of the route due to a low in the Gulf of Alaska that was moving northward. I also told him that in the event that the winds at flight level in the Middleton-Elmendorf sector were stronger than forecast, turbulence would probably be encountered in that area. At approximately 1300P a pilot came in with a clearance for the C-124 to Elmendorf. I told him that his cross-section had not been made up and he stated that he had the upper-section and had been briefed. I then filled out his clearance with the latest terminal and alternate weather, discussed terminal weather with the pilot, and signed the clearance. As the pilot was leaving, Sergeant Holcomb came in. I asked him if he had given the -124 pilot his cross-section and he said it had not been made up and that the pilot would have to come back for it when he tried to clear through MATS Operations. Sergeant Holcomb then went to work on the 1830Z surface chart and the C-124 cross-section. I returned to work on the upper-air charts.

Robert W Evans
ROBERT W EVANS
M/Sgt, USAF
Forecaster

DETACHMENT 4, 4TH WEATHER SQUADRON
McChord Air Force Base
Tacoma, Washington

26 November 1952

STATEMENT

I began work at the MATS Weather Station at 0800P, 22 November 1952. After orienting myself on local weather, the synoptic situation, etc., I began analyzing the 1230Z surface chart. This was completed at approximately 0930P. I then studied the local situation again and also the Alaskan and enroute weather. Three (3) flight folders were to be prepared for the following flights: RCAF North Star to Elmendorf at 1400P; C-54 to Kodiak at 1415P; C-124 to Elmendorf at 1500P. At approximately 1030P to cross-section for the RCAF North Star was begun (legend, etc.). At approximately 1050P I began sketching the 1500Z 700mb chart in order to forecast winds for the departures. At approximately 1125P I had completed the forecast winds for the two routes. I then continued with the RCAF North Star cross-section completing it at 1200P when M/Sgt Evans came on duty. After giving him a brief on the local forecast, I began preparing the C-54 cross-section to Kodiak inasmuch as I would not have time to complete it after lunch in time for their scheduled departure. This cross-section was completed at approximately 1230P. I then made the skeleton form for the C-124 cross-section (legend, etc) so that it could be completed in a minimum time after I returned from lunch. I then briefed Sergeant Evans on the terminal and enroute weather for Military Airways for the RCAF North Star flight. This briefing was as follows: from the forecast position of the low center and the frontal system as indicated on the 1230Z surface chart, solid instrument conditions would be encountered from 140° W to Elmendorf. I expected icing conditions across the system as indicated on the cross-section for the same portion of the route. The terminal weather at Elmendorf was expected to be ceiling 4 to 5 thousand feet, visibility good with no precipitation. At Ladd (the alternate) conditions to be 12 thousand feet broken sky, visibility 10 miles or better with very little possibility of fog. I told Sergeant Evans that that was probably the only one he would have to contend with and that I would be back in time to brief the others. I then went to lunch at 1245P. I returned from lunch at 1340P. As I came in the back door Sergeant Evans was signing a clearance form and by the time I had hung up my hat and coat, the officer whom he had briefed had gone into MATS Operations. When I got over to the map table to start work, Sergeant Evans asked me if I had given the C-124 crew their cross-section. I stated that I had not because it had not been made up. He then said that the officer told him that he had already gotten his cross-section and briefing. I then said that he would be back when he tried to clear through MATS Operations. I then proceeded to sketch the 1830Z chart for the Gulf of Alaska

After reading these two statements, I was not sure if what I was reading could be correct. How does a plane prepare for takeoff, get on the runway, and not have the proper paperwork to do so. Did they receive a copy of the chart plans? In my opinion, while they were doing forecasting reports for three different flights and trying to take their lunch, there could have been a mix up somewhere. Plus, if there was an error in the forecast like the military stated, I can understand the pilot and navigator being off course and crashing into Mount Gannett.

USAF Findings
The board having carefully considered the evidence before it finds:
1. The aircraft crashed in to the side of Mount Gannett around its flight altitude of 9k feet. At which time it was about 30 mile off course.
2. There was no indication of malfunction of mechanical or radio equipment.
3. The most probable cause of the accident was a navigational error attributed to the pilot
4. A contributing cause of the accident is that the winds were incorrectly forecast.
5. A probable contributing cause was precipitation static which made reception impossible.
6. It is a probable conclusion that the aircraft crashed prior to its ETA at Whittier.

Actually, it was much more than that.

During the same timeframe, there were several military planes that went missing. This would include nine US military planes that crashed or vanished during this very short timeframe. Those crashes/vanishings totaled 288 lives, and the questions being raised were why were these planes crashing and what could be done to fix this. The first week in January of 1953, the House Armed Services Committee summoned General H.S. Vandenberg and other top Air Force officials demanding answers about the 86 servicemen that were killed on Dec 20, the 52 servicemen that died Nov 22, and the other seven planes crashed. I have one question; why didn't the military take the proper actions prior to losing 288 lives? The Aviation Safety database has all the data on the plane crashes and fatalities on their website:

https://aviation-safety.net/database/types/Douglas-C-124-Globemaster/database

The C-124 had a variety of problems associated with its anti-icing equipment, autopilot, brakes, and instrument visibility. Until Wright Air Development Center (WADC) engineers could devise a solution to ice

formation, pilots were simply told to avoid icy conditions. At the end of 1952, all C-124s of the 22nd Troop Carrier Squadron were grounded because of fuel tank leaks. In early February of 1953, after fuel cell modifications, the big planes returned to the skies. In July 1953, a number of C-124s were grounded again pending inspection of their engines after a number of engine fires. On June 18th, 1953, the worst air disaster of that time occurred at Tachikawa Air Base in Japan when an engine fire caused the crash of a C-124 shortly after takeoff, killing the 129 passengers aboard. Some of the planes were returned to service the following month, but many remained grounded till the end of the war, awaiting new generators. Despite its problems, the C-124 had demonstrated that it was the cheapest air transport per ton-mile (one ton of freight carried one mile, as a unit of traffic) in the Air Force inventory.
(Source: http://www.globalsecurity.org/military/systems/aircraft/c-124.htm)

If the board opinion states the accident was caused by an unforecasted increase in the winds which the pilot encountered in the Middleton Island area and by poor radio reception from severe

Air Chief Called In Crash Inquiry

House Group To Quiz Vandenberg, Aides

WASHINGTON, Jan. 6 (*P*)— Gen. Hoyt S. Vandenberg and other top air force officials were summoned before the House armed services committee today to explain, if they can, the causes of recent military crashes.

The open hearing signaled the first of what seems certain to be a deluge of investigations by the new Republican Congress.

A total of 288 lives were lost when nine American military planes crashed or disappeared during November and December in this country and the Far East.

Chairman Dewey Short (R., Mo.) asked General Vandenberg, air force chief of staff, and his aides to provide the house group a full explanation, including an estimate of whether the causes were mechanical failures or pilot errors.

"We want to know the reasons, so that we can prevent future disasters," Mr. Short told a reporter.

The hearing, a preliminary one, will be concluded today. Mr. Short said the committee then will decide whether to schedule a full inquiry.

Eighty-six servicemen were killed Dec. 20 in the crash of a holiday-bound C-124 Globemaster near Seattle. A month earlier another C-124 disappeared off the central Alaskan coast with 52 persons aboard.

Seven other military disasters in November involved a navy patrol bomber, four C-119 "Flying Boxcars" and two smaller transport planes, a C-54 and a C-46.

precipitation static, which is a known to have existed in that area at the time. Then how could it be Pilot and Navigator error?

Col. Jack Stovall felt the same why. He stated: the finding that the most probable cause of the accident was due to navigation error. The term "error" implies that a mistake was made by the pilot when conceivably such action was avoidable. This is not considered quite accurate.

If the Board considered that an error was made, then the recommendations should include desired corrective action in an effort to prelude the reoccurrence of such an error. If, under the conditions existing at the time, a logical technique for preventing the accident was not available, then the cause of the accident should be attribute to inadequate equipment rather than pilot error. ~United State Air Force accident report.

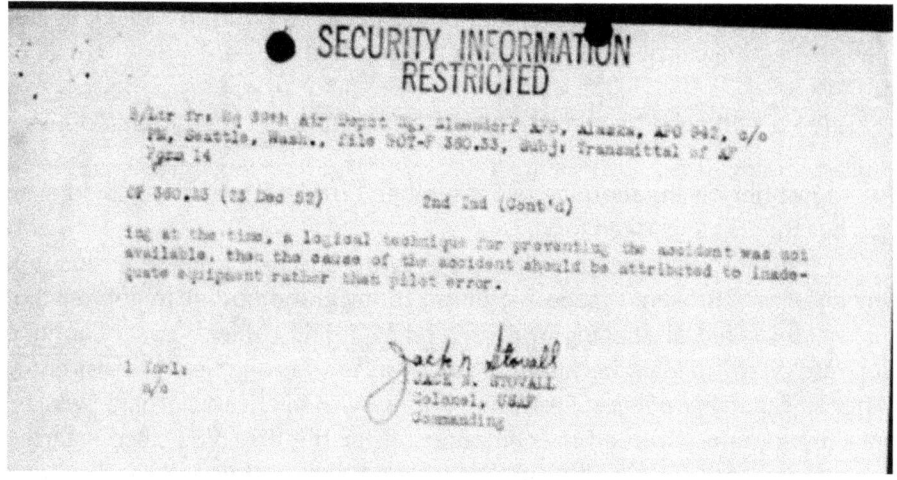

To allow the families to feel that the pilot and navigator were in the ones in error for all of these years is, for me,...so wrong.

After all of this research, and reading the report, I needed to bring my grandmother up to date on where things were and what I had discovered. She was now in the hospital having tests done. Her health had been slowing deteriorating over the years. However, we weren't really sure if it was just old age or if there was anything more serious going on. "Hello Grandma, how are you doing? "

"I'm doing well baby, any news yet?"

"No, Grandma, no good news yet. I've written so many letters, and I don't know how to turn this anger around. They left them. They just left them there and never went back. What ever happened to never leaving the fallen behind? I guess it costs too much."

"Child I know you will figure it out eventually. I have faith in you. It has been over 40+ years since his plane went missing. Answers will not happen overnight just stay on them.

I spent the next couple of hours sitting by her bedside and talking with her about my quest to find out what happened to the plane carrying these lost soldiers.

I would go back almost every day for a week to update her on everything. While she was dozing in and out of sleep, I kept talking to her. At one point, I stopped, and she opened one eye to see if I was still there.

"Please keep talking honey, I want to hear everything."

Deep down inside, I knew her health was failing, and that she may not be here once I finally uncovered the truth about her husband. She was getting worse and had to go into surgery. During the surgery, they discovered that she had cancer. This made me want to get the flag for her as quickly as I could. I had to complete my mission before my grandmother left this earth.

On September 11th, 2001, less than one week after me talking by her bedside, I got a call at work that saddened me to the core.

Tonja, you need to go to the hospital right now. It's your grandmother.

I lost her on the morning of September 12th, 2001, before I could even get her the flag she asked for.

I mustered up enough willpower to write one or two more letters, but my heart was broken. I made a promise to my grandmother to get her that flag, and I failed at keeping it. At this point, I didn't think I could continue forward on my mission because part of it was for her, and now she was gone. After saying my goodbyes, I returned home. With my heart hurting, I walked into my room and turned my computer off for the first time since I started looking for answers. I packed up the letters I had written and received and pulled down the maps from my wall, vowing to never bother the military again about the plane crash.

Almost a year went by before I would speak of or do anything about the plane crash again. My father stopped over to check on me because he and my husband were worried about how I just gave up so abruptly. My husband said to my father, "A part of the quest died when Ms. Dot passed away, and I don't know what to do to get her back into things."

My father came into the room and asked if he could talk with me for a moment.

"Tonja, I'm still here. I'd like to know what happen. Even though I said I didn't want to know, I do. Yes, at first I didn't want you to do this because I didn't want you to open something up I wasn't ready for, but I'm ready now."

I spent several days thinking about what he said and wondered if this was something I even wanted to continue. My husband had come into my office, where all the research, phone calls, and letter writing had been done. I stood there in the middle of the floor, looking around at all the stuff I had boxed up.

"Tonja, I've got a package from the military addressed to you."

I turned and looked at the box in his hands. All of a sudden goose bumps grew all over my arms. An excitement started to build up inside of me again, and a smile slowly crept on my face. I grabbed it out of his hands with the excitement of a child in a candy store. I ripped it open and peered inside the box.

Chapter 4 - Second Chance

Inside the box was a tape and some letters about the crash. I cried because I had lost my grandmother, I had given up, and I never thought the day would come that the Air Force would go back just for little 'old me. But they did.

DEPARTMENT OF THE AIR FORCE
HEADQUARTERS AIR FORCE SERVICES AGENCY

18 October 2002

Mr. Marshall A. Blair
Air Force Mortuary Affairs Division
9504 IH 35 North, Suite 320
San Antonio TX 78233-6635

Ms. Tonja Anderson
7210 N. Manhattan Ave., #2121
Tampa FL 33614

Dear Ms Anderson

On 10 September 2002, Mr. Huey of this office over flew the mishap site of the C-124 in which your grandfather, Airman Basic Isaac W. Anderson, Sr. died. The grid location for the site was provided by the Rescue Coordination Center as North 61' 12" and West 148' 10". Several passes were made over the area by the helicopter. No wreckage is visible on the mountain or the glacier. A video of the area was made and is forwarded for your review and retention. Additionally paper copies of the photographs of the mountain taken by the Dr. Moore and LT Sullivan are provided. The video duplicates the mountain peak as depicted in the 1952 photographs. Attached is a color map of the area with Mount Gannett indicated by the red pointer is provided. At the present time there is no estimate as to the depth of the snow and the glacier.

Mr. Huey briefed the Central Joint Mortuary Affairs Office on the over flight on 8 October 2002. Representatives of the Army Memorial Affairs Office and the Navy Decedent Affairs Office were present. Each was provided copies of the video, paper copies of the 1952 photographs and the map. Safety of the recovery team is of paramount concern in mishap site recovery operations. All were in agreement that the site is not recoverable.

We hope that the material provided will provide some comfort to you and other members of Airman Basic Anderson's family. If we can be of further assistance, please contact Mr. Huey by calling 1-800-531-5803, 8:00 AM to 4:00 PM (CT) Monday through Friday.

Sincerely

MARSHALL A. BLAIR
Chief, Mortuary Affairs Division

C-124 Mishap Site
Over Flight
10 September 2002

After I received the tape, someone from MacDill Air Force Base called me about my grandfather. We spoke, and they arranged for a service on the base, just like my grandmother had wanted. I couldn't believe we were finally getting him a proper funeral. We were finally getting our flag. The flag he so rightly deserved. I was so excited I called my father right away since he had encouraged me to keep moving forward. I could hear him choke up with tears over the phone. He had become just as emotionally invested in the journey as I. We agreed on holding a service at the MacDill Air Force Base in Tampa, Florida.

We arrived at the base; the chaplain gave a brief service, and we were then directed outside to the lawn. A bugler and soldiers with rifles were on the grass. My father, his wife, his children and his grandchildren, all lined up and stood as the bugler played "Taps." We all looked at each other as if this moment was not happening.

One of my children leaned in and asked, "Mom, is this for the man you did all that work for?"

I replied, "Yes…Yes, it is. I wish grandma were here."

At the end of "Taps," the gunmen turned, and the 21-gun salute was performed. Shot after shot caused us to jump. Tears ran down my face because this was a bittersweet moment for me. I'd been able to accomplish what I'd set out to do, but my grandmother was not here to see it. We all walked back into the church, and the folding of the flag ceremony started. The soldier walked up to dad got down on one knee, whispered some words and the flag was presented to my father.

DEPARTMENT OF THE AIR FORCE
6TH AIR MOBILITY WING (AMC)
MACDILL AIR FORCE BASE, FLORIDA

This Memorial Folder
is presented to you as a
cherished memory of your loved one,
may it be kept near forever.
We present this to you with deepest compassion
and our sincerest expression of sympathy.

AB Isaac W. Anderson

The MacDill Air Force Base Honors and Ceremonies Team
listed below, provided honors for this memorial service

MEMBER IN CHARGE

MSgt. Mulqueen

HONOR GUARD MEMBERS

TSgt Douglas	A1C Ward
SSgt Sanborn	A1C Donahoe
SSgt Cannon	SrA Leggette-Counts
SSgt Long	SSgt Felder
SrA Valentine	SrA Rodriguez
SrA Provost	SrA Miller
SrA Moseley	A1C Lacombe
A1C Schneider	SrA Burrage
SrA Jones	A1C Dirsh

Taps played by:
SrA Casey-Olliges

Please refer any comments to MacDill Air Force Base
Honors and Ceremonies at (813) 828-5191 or write to:
Brigadier General William W. Hodges
Commander, 6th Air Mobility Wing
8208 Hangar Loop Dr.
MacDill Air Force Base, Florida 33621-5541

AMC--GLOBAL REACH FOR AMERICA

Chapter 5 - Making it Through the Fire

Life has unexpected moments, and they push you back to your soul's mission. After the funeral for my grandfather, I put the plane crash aside. My goal originally was to get the flag for my grandmother. I had accomplished that goal. I felt satisfied. My family was ecstatic, and I thought my journey digging through the past was complete. I packed up all my documents, letters, and paperwork and put them in boxes. Stored them away in a wooden cabinet that belonged to my grandmother and placed the flag, encased in glass, on top. I stood there for a moment and let out a long breath of air and stared at my little shrine of truth.

I moved on with my life doing my normal daily tasks. However, as time drew on a nagging voice kept me up at night. I had helped my grandfather's legacy, but what about the rest of the 51 families? It's not fair for them to not have closure. They probably did not know any of this information that I had gathered. How could I help them as well or even find the other family members? I wanted to help, but wasn't sure if I could.

Then in January of 2003, a year and a half after my grandma's death and MacDill Air Force Base giving me the flag; there was a knock at our door. My husband answered to see a frantic lady stating that her apartment above us was on fire. He went upstairs to help, but the fire was out of control. He came back to get our twins and me, and we watched our apartment go up in flames. It all happened so fast we had no time to grab any possessions. Then I started to panic. My grandfather's flag! I worked so hard to get this for my family, and now it's going to be burned in a senseless fire along with all my documentation. No this can't be happening. My files! His flag cannot be lost in a fire. It cannot !!!

The fireman came over to us and stated that our building was unstable and we could not go in, but they would try to see if anything of salvageable.

I looked the fireman in the eye and stated, "In the first bedroom on the left is a flag in a case with items around it. If it's still there, that's all we want." My husband looked at me and said, "Mmm, you didn't ask for my wallet or your purse?" I looked at him and just smiled and said sorry.

Through the smoke and water, the fireman came over to me and in his hands was the flag in its case. He looked at me and said, "Ma'am the fire burned up and all around the wooden case, but didn't touch anything on or in it."

I gave him a kiss and a hug, saying, "Thank you, sir!"

Later that night, my husband and I went back to the apartment to retrieve some of our personal items. He went into our bedroom, and I headed straight to the room where I'd done all my work. I looked into the wood case, and I found the box that held all my research, including the microfilm and videotape sent to me by the military, untouched. Goosebumps grew all over my arms once again. I couldn't believe it. The line of the fire went right up and over the wooden case and the flag without touching it at all. My stomach started to get queasy. It was as if something was there protecting all my research. It was a sign. A sign from the universe that I need to keep pushing forward on my mission.

Months after the fire and almost losing all my research, I received an email. It was from a family member of Airman Howard Martin, asking if I was the same Tonja Anderson from the internet.

I wrote, *Mr. Williams, if I may ask, how you did find me.*

His reply was:

To: Tonja F. Anderson
Subject: RE: Air-Crash C-124

Tonja,

The way I found you was that you made a post (maybe to an Alaskan USAF group on Kodiak Island) on January 26, 2001, concerning your grandfather and the C-124 crash. I discovered this post when I was doing some deep searching on the web. I printed the post and filed it in a loose leaf binder I keep concerning this issue. At various times in the past, I've attempted to find you with no luck. Then I tryed Facebook.

Mike

Subject: Looking for Information

I'm looking for information about a unit. My grandfather was in the Air Force, he died in a C-124 aircraft accident on Mount Gannett Alaska on Nov. 22, 1952. He was in the 625th AC&W Sq. 10th Air Div(Def), Apo 942. Can you tell anything about the unit? I'm doing a search about his death and trying to get the Airforce to do a search and recovery of this plane and its remains. I know this going to be hard, but I need to know what happen to him and the other 51 people aboard this plane. If anyone can help, I'm grateful. Thank you, Tonja Anderson

We exchanged emails and information for years after this. It was great to find someone out there who was looking for that same thing I was, answers. I had put the crash and the investigation on the back burner because I had lost contact with Ms. Patti Frigillana from Hawaii (her father was Major William Jackson, I had received the flag for my grandmother and—somewhat—closure. I say somewhat because I still didn't understand why the military was right there and had seen specific items, but had never brought any of them back.

Airman Howard Martin, his siblings, James, Ray, Kay, Max, and Paul, Fran and Kay

Mike sent me a copy of a letter he'd received from JPAC, and it really caught me off guard because I knew this statement was incorrect.

JOINT POW/MIA ACCOUNTING COMMAND
310 WORCHESTER AVENUE
HICKAM AFB, HAWAII 96853-5530

DCO (ER)
2008-112
16 April 2009

Dear Mr. Williams,

Thank you for your letter of 12 December 2008 concerning the "Archaeological Site Report For" of a USAF crash site on Mount Gannett located in Alaska which you sent to JPAC ADM Crisp asked me to provide you with your requested information on her behalf.

This crash site had not been previously reported to JPAC or its predecessor, the US Army Central Identification Laboratory (CILHI). Our historians have established a case file for this C-124 crash site and assigned Incident Number OTHER 27-Internal (OTH27-J) to the case. The maps, selections or the aircraft accident investigation and a couple of photographs you included with the site report form are also incorporated in the case file. The basic circumstances of the loss incident are clear from the materials attached to the JPAC Site Report Form. The records suggest that no remain were recovered after the incident due to the severity of the field conditions and the terrain. We have requested additional records from the Air Force Historical Research Agency and the Washington National Records Center to assist us in evaluating the possible course of action to resolve this loss. Based on the information received from you, we are tracking this loss incident as an open case pending further review of the complete records. Preliminary review of the data concerning the loss indicates that a JPAC high altitude team could possibly reach the location and conduct an investigation of the site.

Thank you again for your support in JPAC's mission for the fullest possible accounting. If you have additional questions or information, please do not hesitate to contact me at the above address or via email
Sincerely,
Deputy to the Commander for Public Relations and legislative Affairs

In November of 2000, I received a letter from Department of the Air Force, Headquarters, United States Air Force, Washington DC. It was a reply to an email I'd sent to the Defense Link, concerning the recovery of the wreckage from my grandfather's aircraft on Mt. Gannett in Alaska.

Colonel Kelly stated that they contacted the CILHI and that they were already aware of this accident and would be responsible for initiating the recovery operation. You may want to discuss your concerns with them by contacting them directly at: CDR USA CILHI/TAPC-PED-H, ATTN: Dr. Tom Holland, Building 45, 310 Worcester Ave, Hickman AFB, HI 91853-5530.

DEPARTMENT OF THE AIR FORCE
HEADQUARTERS UNITED STATES AIR FORCE
WASHINGTON D.C.

HQ USAF/ILV
1770 Air Force Pentagon
Washington, DC 20330-1770

0 2 NOV 2000

Ms. Tonja Anderson
7210 N. Manhattan Ave, #2121
Tampa, FL 33614

Dear Ms. Anderson

This is in reply to your DEFENSE LINK e-mail concerning the recovery of the wreckage from your grandfather's aircraft on Mt. Garrett in Alaska.

Mr. Gary Huey from the Air Force Mortuary Office has spoken with you recently. As he mentioned, there are currently no plans to conduct any recovery operations. Since this crash also included members from the Army and the Navy, our mortuary personnel have contacted those branches to determine if they have any plans for recovery. Unfortunately they do not.

We have also contacted the Central Identification Laboratory, Hawaii (CILHI) who is aware of this accident and would be responsible to initiate the recovery operation. You may want to discuss your concerns with them, by contacting them directly at:

CDR USA CILHI/TAPC-PED-H, ATTN: Dr. Tom Holland
Building 45, 310 Worcester Ave
Hickam AFB, HI 91853-5530
Telephone: (808) 448-8062 x 151

Should any new information become available, our mortuary staff will contact you.

Sincerely

MICHAEL J. KELLY, Colonel, USAF
Deputy Director of Services
DCS/Installations & Logistics

From: hickam-cilhi.army.mil>
To: <t.felisha@gte.net>
Cc: cilhi.army.mil>
Sent: Friday, August 17, 2001 1:31 PM
Subject: RE: C-124

Tonja — it is good to hear from you, it has been quite a while since we communicated (negligence on my part!). I have taken a new position with the organization and am out of the research department. I have cc'd Mr. _____ who is assuming my role for the interim. Unfortunately, this email will blind-side him as I do not think this is one of the issues I had discussed with him during the quick transition! The last action I recall on the case was we were still after the Air Force to get us a manifest of all passengers and crew so that we could start rounding out our data base. The stopper as I remember it was the AF was somewhat skeptical as far as getting the State Medical Examiner involved in the case, as per Alaska State law, the ME has responsibility, even in the case of Military Airlift incidents. Need to due a follow up. Please bare with us while I back brief Bill and allow him to establish contact with the Air Force.

Once again, it's another — we're working it — please be patient — answers. We really are concerned about your grandfather's incident, as well as the other service members and crew, we can only hope we can play this through with AF and ME involvement.

Other wise, hope all is well with you, and one of us will be giving you an update. Aloha -

Deputy Operations Officer
USA CILHI, Hickam AFB, HI 96853
(808) 449-5260 ext 211

This had me wondering which statement was true—the letter sent to me in 2000, or the one sent to Mike in 2009. So, I pulled out my box to see if I had any additional letters and I did. In August of 2001, I wrote to Mr. Huston, and he replied via email providing me with an update on his transition into his new position and putt me in contact with his replacement. He also went on with giving me an update on where things were with the C-124. CILHI was still after the Air Force to get them a manifest of all passengers and crew so they could round out their database. The stopper was that the Air Force was somewhat skeptical as far as getting the State Medical Examiner involved in the case, as per Alaska State Law, the ME has the responsibility, even in the case of the Military Airlift accidents.

Also in January of 2009, Joint POW/MIA Accounting Command, J2—Mr. Chris McDermott, 310 Worchester Ave, Bldg. 45, Hickman, AFB, HI 96853-5530 from AFHRA/RSA, 600 Chennault Circle, Maxwell, AL 36112-6424. The subject of this letter was Accident Report 52-11-22-8. On the very top of the paper, hand written was ALASKA OTHER JPAC INCIDENT 27. Ms. O'Connor sent Mr. McDermott a copy of the accident report he needed as he was doing some research.

This is where I think the ball was dropped by JPAC/CILHI or that their focus was more on recoveries overseas. It should not matter if the servicemen lost their lives in another country or right here in the United States. These men died for their country.

Putting those thoughts on the backburner, Mike and I wondered if the C-124 had moved. Where it was located now and if any other parts of the plane had surfaced. Mike hired a photographer, Calvin Hall in Alaska, to do a flyover of Mount Gannett and the surrounding area. He provided Calvin with copies of the crash site with background images to help him gauge the exact area for his flight.

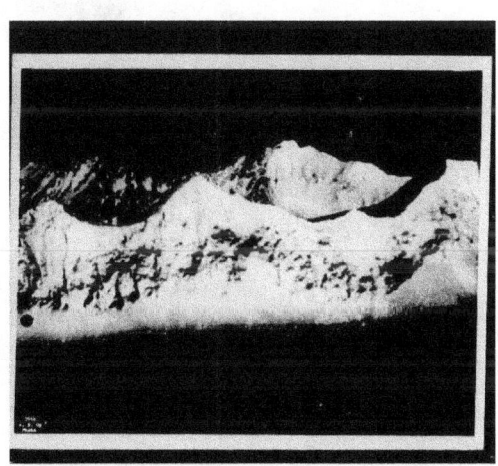

Calvin reached out to Mike on his journey to Mount Gannett and the crash site. In his email, he stated:

The actual crash site, found by the photo reference is not on the West ridge, but on a "small" ridge that goes South and then Southwest. Because I thought it was on the West ridge I had been trying to fly it in late afternoon to have nice sidelight to show detail. I ended up being at the crash site in total dark shadow instead. With the good optics and digital camera I was still able to get good shots. While flying over upper Colony glacier I noticed a spec with movement. It looked close by, but I could just barely make out that it was a helicopter parked down on the glacier. With my 200 mm lens I was able to see the 4 men down there. I could not see any signs of aircraft debris from the crash. The only possibility would be on the rock wall. With a helicopter that could hover at that altitude, and shot in morning sunlight, would be about the only way to spot anything. The glacier at that elevation never melts off the top snow from the previous winter, so it would be extremely unlikely to find anything unless it has been transported miles down glacier. If there was a more extreme summer melt back, there would be a very slight chance of finding debris at the top edge of the glacier on the rock wall. ~ Calvin

Mike and I looked for other family members as we did our search together. We were contacted by a Mr. Jones in 2011. He informed us that he was part of several search crews that went back to look for the C-124 crash in 1952 and 1953. This was exciting because he was the first person I'd spoken to with personal information, and I was not aware that there had been several recovery missions.

Aircraft Crash on Mt. Gannett Alaska November 1952

A USAF Aircraft C-124 tail number 1107 crashed on Mt Gannett November 24th 1952 The Aircraft was reported missing at 9:45 P.M. PST There were 52 Men on board. There Were No Survivors. The wreckage was found a few days later on Mt Gannett.

My name is Robert M. Jones. In November 1952 I was a USAF Para-Rescue jumper attached to the 71st Air Rescue Squadron 10th Rescue Group stationed at Elmendorf AFB Anchorage Alaska

I was part of a team assigned to a recovery mission to try to recover whatever remains or other information we might find at the crash site. There were three missions dedicated to this crash from 1952 to 1954 that I was either aware of our part off.

Mission One

On 28th November 1952 this mission was carried out by a Lt. Thomas Sullivan and Dr. Terris Moore who at that time was the President of the University of Alaska Dr Moore was the owner of a Piper Super Cup on skis. Dr Moore volunteered his time and aircraft to help with this mission. Lt. Sullivan was assigned to go with Dr. Moore and to try to land their aircraft as close to the crash as possible and to confirm that this was the missing aircraft, and that there were no survivors. They did reach the crash and confirmed that this was the missing aircraft and there were no survivors. This also was confirmed by a statement by Lt. Sullivan who was a member of the Inspector Generals Special Investigations Squadron at Elmendorf AFB Alaska.

Mission Two

I was personally involved in this mission, however I cannot remember the date it was carried out.. I believe it was the early part of 1953. This was a rather large mission involving the US Coast Guard Cutter Storis. The cutter was brought up from its station in Juneau Alaska. A large barge was towed by the Cutter, and two or three H-5 helicopters were placed on the barge. The copters were from the 10 Rescue. These helicopters were limited in their ability to hover at high altitude(around 4000 feet or so). Our team, I believe was 6 men and 2 officers. The ship pulling the barge was then moved to Prince William Sound, and to a small fjord, I think called Port Wells, located on the water side of Mt. Gannett. The plan was to copter us as high up the mountain as the copter could hover.—and, to drop us in the snow for the attempt to go over from this side of Mt. Gannett and to the crash site. The weather was bad for the first few days. Finally there was a break, and the 2 officers made an attempt. We were to wait for their orders to follow if the mission was feasible. The attempt failed. The conditions were too unsafe to continue—so the mission was called off. We all returned to base.

NOTE: I don't remember the name of the officers. I believe they were from the IG special Investigations Squadron.

Mission Three

This also was a fairly large undertaking. After the failure of mission 2 we were told to prepare our equipment for another attempt in late Summer of 1953. The powers that be hoped that the Summer weather would thaw the ice and snow—and that we would have greater access to the crash scene. Here again, I cannot remember the exact date of this mission. I do know it was late Summer or early Fall. This mission started by bringing in a more powerful helicopter from the Southern United States. This Copter H-19 was disassembled enough to fit into a C-124, and was then flown into Elmendorf Air Force Base. There it was reassembled, and readied to take our Team as close as possible to the crash site. This Copter was flown to the Palmer, Alaska Airport, and used to take us, two men at a time as high up the mountain as they could hover. This ended up to be on Surprise Glacier at about 1000 feet from our camp site on the snow bowl in front of Mt Gannett.

Mission Three Page Two

Our 10th Rescue aircraft air dropped our equipment. Camp was set up the first day. The next day the weather was sunny and warm. We finished setting up camp. We then made our first attempt to get to the crash site. I would guess our camp was 1000 feet or less from the mountain face. We hiked to the bottom of the face of the mountain, and found that there was a large crevasse between the mountain and the ice and snowfield we were standing on. As we looked up on the mountain face, we didn't see any large items from the wreck. No engines, tail parts, or fuselage parts. Some of the men thought they saw some evidence of metal parts high up towards the top of the mountain face. I don't recall seeing anything. As we were looking for evidence, large rocks started to fall around us. A small rock hit one of the sergeants on the head, so it was decided to retreat back to the camp. We planned to radio in to headquarters for more equipment we would need. Ice ladders, helmets, and rock-climbing gear we would need if we were to proceed. Due to the warm weather, the crevasse and steepness of the rock face, this in my opinion it would be to difficult to continue. As the afternoon proceeded, the weather changed, and it began to snow. By the next morning, it turned into blizzard conditions. It snowed for days, and kept us all in our tents. We did on occasion have to go out and shovel snow off our tents to keep them from collapsing. It was obvious to all of us that we were too late in the season—Winter had arrived. It was now a matter of getting us off the mountain at the first break in the weather. This came after a few days. We were told that the helicopter that had brought us up to the mountain, had a broken transmission, and that we would have to walk out. It took a day to get to the bottom of Surprise Glacier. We then had to spend the night on a rock side of a mountain next to Kink Glacier. The next day our own H-5 Helicopters from 10th Rescue picked us up in the valley, and took us to the Palmer Airport. We were then trucked to the Elmendorf Air force Base. This was the end of Mission 3.

Comments

First let me say, anyone interested in this information is welcome to use it in any way they see fit. This was 59 years ago and I cannot remember the dates of the events, but the events took places as described. It was saddening to all of us that Mission 2 and 3 couldn't have been more successful. Sometimes the weather can beat the best of men and plans.

Now for the mystery. Why was the Air Force so bound and determined to reach this wreck? Lt. Sullivan stated in the first Mission that he felt it wasn't practical to further explore the crash site. He noted the remoteness and snow and ice at the site, and very little left of the wreck. And why were the IG investigators assigned to each Mission? (Inspector General) On Mission 3 there were 2 Majors, a Major Frittier, head of Operations 10th Rescue; the other Major Crepeau of the IG investigation Squadron. Both men were on the lead team of the Mission. Was there something special on that aircraft that they felt they must try to retrieve? The scuttlebutt from the team was maybe there was payroll for some of the Squadrons at Elmendorf Air Force Base. Could it have been secret documents? Or just a good training event. Also the Korean war was on at this time. I guess I will never know.

On Mission 3 we spent such little time at the Mountain face and, saw so little. I often wondered if we were at the wrong place on the mountain. As I recall, I saw nothing; others said they may have seen small parts. And Lt. Sullivan in his report, said they walked over to the crash site from Surprise Glacier from where they had landed. Could we have been on the wrong side of the mountain, or too far up to find the wreckage? The officers that planned the Mission and had the whole summer to figure it out, and had us concentrate on the area that I described. Could most of all that was left of the Airplane have fallen into the crevasse that I had described earlier? I believe that by now other people have been to the site and it would be interesting to see their information. At any rate this was a tragic event. My heart goes out to all of the families and friends of the victims. And, a last note, we at 10th Rescue, also had a good friend on this flight—Captain Ponikvar, a navigator who also was in charge of our Rescue Team. He was returning to Base as a passenger on this flight. May they all rest in peace.   By Robert M. Jones Jan 28, 2011

For many years of my research, I'd thought there were no other search missions. I later got a copy of the 10th Air Rescue Group report on July-December 1952. On pages 5 and 6, they reported on Dr. Moore's mission on December 7th, when a party of six was dropped off on Surprise Glacier to search for the plane with no success due to heavy snow cover, and a 10-day land rescue yielded nothing either. After reading this, it still made me wonder why they never grabbed the items on top of the crash site while they were there during the first landing. We research for many years together and still nothing.

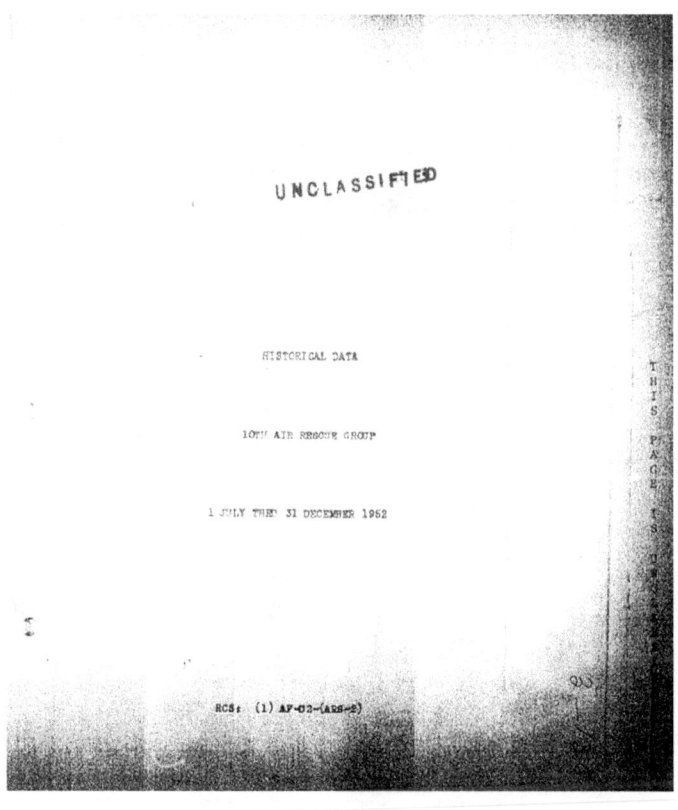

History of 10th Air Rescue Group fr 1 Jul 52 thru 31 Dec 52 (Cont'd)

the top of Mt. Silverthrone.

Approximately one week later a second C-119 aircraft disappeared in the vicinity of Kenai while enroute from Elmendorf to Kodiak. A maximum effort search was made largely under adverse weather conditions and the aircraft with its 20 passengers was never located. Approximately eight feet of snow and heavy slides occurred in the area of highest probability shortly after the aircraft was missing. The search was suspended four weeks later.

This series of tragedies reached its peak in the disappearance of a C-124 Globemaster enroute from McChord AFB, Washington to Anchorage with 52 persons aboard 22 Nov 52. Last reported over Middleton Island, the aircraft had only 140 miles to go and a 9000' range to clear before reaching its destination. A maximum effort search was immediately instituted and on the third day, the first day in which the weather broke in the vital area, wreckage was sighted at the 8300' level near Mt. Gannett on Surprise Glacier. The small visible remains of the aircraft were in a condition that did not permit positive identification from the air and there was no sign of life. Due to the altitude and terrain features, landing at the scene was not feasible. Doctor Terris Moore, president of the University of Alaska, proceeded from Fairbanks in his ski-equipped Super Cub modified for high altitude landings, and the following day, after landing at the accident site, positively identified the wreckage

---

History of 10th Air Rescue Group fr 1 Jul thru 31 Dec 52 (Cont'd)

the H-5's piloted by Captains E. A. Ray and P. E. Kimberling were able to transport six of their party to a base camp at the 5500' level on Surprise Glacier. The party was supplied by air drop from Elmendorf and reached the crash site at approximately the 8300' level on 7 December. Two days later all personnel were returned to Elmendorf from the advance base by H-5's in a shuttling operation. Landings and take-offs at this altitude (5500') were extremely hazardous and represented performance considered exceptional for this type aircraft. No identification or evacuation of bodies or cargo was possible at this time due to terrain features, altitude and almost complete coverage of the wreckage by heavy snow and snowslides. This operation has demonstrated extremes in several phases of arctic search and rescue in the sighting of so small a piece of the wreckage which was visible, the ten day land-rescue expedition, the helicopter support in extreme weather and at extreme altitude, the hazards of the climb by the land party up the glacier to over the 8300' level, and the hazardous landing and take-off performed by the Super Cub. All of these were performed in extremely low temperatures.

A usual number of other search and rescue missions were performed during this period, none particularly noteworthy except perhaps the rescue of two officers who crash landed a fighter aircraft on a frozen lake near Iliamna and who were rescued by an SA-16. This was our first emergency landing rescue and take-off from a frozen surface by use of an SA-16

I will never forget the day I started searching the internet and came across a story by KTUU-TV a television station based out of Alaska on an old plane found in the mountains.

ANCHORAGE, Alaska— Possible military aircraft debris, along with material that may be bone fragments, has been discovered in the Knik Glacier area according to the Alaskan Command.

Alaska Army National Guardsmen on board a UH-60 Blackhawk flying a routine training mission discovered the debris at about 1 p.m. Sunday, and conducted a brief aerial inspection before returning to Joint Base Elmendorf-Richardson.

Lt. Tania Bryan, director of public affairs for the Alaskan Command, said the crash was believed to be that of a vintage aircraft and "not recent."

She says details about the crash are being withheld pending possible notifications of next of kin.

A recovery effort for the wreckage is being considered by the U.S. Joint POW/MIA Accounting Command, which conducts search, recovery and laboratory efforts to locate lost service members.

The Federal Aviation Administration has placed a temporary flight restriction on the area, and aviators are being asked to avoid the vicinity as personnel investigate the site.

A feeling came over me that I couldn't put into words. I was jumping for joy and running through the house while my husband looked at me like I was crazy. He asked what was going on and I told him.

He said, "Honey, you don't even know if it's your grandfather's plane."

I said, "It is...I just have a feeling it is!"

I made a phone call the next day.

"Hello, my name is Tonja Anderson-Dell, and I would like to speak with Mr. ******."

"Please hold, Ms. Anderson."

"Yes Ms. Anderson, how are you? It has been a while."

~we both laughed~

I said, "Yes, it's been a long time. Please tell me is...is...is it the plane?"

"I cannot tell you officially, but unofficially, yes—it's your C-124 Globemaster."

I did everything to hold back the tears as I fell to the floor holding the phone. The first thing I did was call my father

"Daddy, they found the plane."

"What?" He asked.

I repeated, *"They found Granddaddy's plane...they found it!"*
He said, *"Wait, start from the beginning."*

I later sat down and emailed Mr. Williams. He's the family member of Howard Martin, one of the men on the plane. I asked him to give me a call as I wanted to give him this news over the phone and not through email. We would later talk about what I found, and we both got into research mode. When would I get the official call? Wait, my father is next of kin.

Even though we knew that the plane was the missing C-124 Globemaster, we had to wait for the official report from JPAC. The official report was due to come out June 20th or 21st.

POW/MIA TEAM INVESTIGATES AIRCRAFT WRECKAGE IN ALASKA
By | June 20, 2012
Specialized JPAC investigation team searches for evidence of unaccounted-for Americans
JOINT BASE PEARL HARBOR-HICKAM, Hawaii (June 20, 2012) - A specialized investigative team from the U.S. Joint POW/MIA Accounting Command (JPAC) arrived in Alaska yesterday to investigate an apparent aircraft crash site in the Knik Glacier area of Alaska.
On June 10 at approximately 1 p.m. an Alaskan Army National Guard UH-60 Blackhawk helicopter crew discovered what appeared to be an aircraft crash site while conducting a routine training mission.
Following additional search and rescue missions by Joint Task Force-Alaska and the Alaska National Guard at the suspected aircraft crash site, JPAC forward-deployed a five-person team to further survey and assess the site and develop recommendations for potential recovery operations in the future.
With full knowledge and cooperation of local military units and governmental agencies in Alaska, the team will investigate the site for about three days, searching for any evidence that may positively correlate the aircraft wreckage to a known incident.
Falling directly under the U.S. Pacific Command and employing more than 500 joint military and civilian personnel, JPAC continues its search for the more than 83,000 Americans still missing from past conflicts. The ultimate goal of the Joint POW/MIA Accounting Command, and of the agencies involved in returning America's heroes home, is to conduct global search, recovery, and laboratory operations in order to support the Department of Defense's personnel accounting efforts.

I called JPAC to speak with Captain Jamie Dobson to find out why it was not confirmed on the 20th as we were told. She informed me that it could take anywhere from six months to six years before this plane could be identified as the missing C-124 Globemaster. I thought I was about to lose my mind when I heard those words. Why would it take that long when they'd picked up several items from the glacier that should allow it to be easily identified?

I picked up the phone and made some calls because I knew this was wrong.

While searching the internet again for more updates, I came across photos posted by JPAC on Flickr. Captain Jamie Dobson posted the photos of the team on Colony Glacier and labeled them as the "missing C-124 Globemaster." After seeing these photos, I was enraged. I'd just spoken with her the day before, and she told me they didn't know if it was the plane. But then she posted photos bragging all over the internet! That was a slap in the face.

I shared these photos on the C-124 Facebook page I started February of 2010 to allow the families to get an insight as to what things looked like on the glacier. The more upset I got I decided to create an account and comment on the photos. I posted, "I want to thank you for posting these photos and PUBLICLY announcing that it was the missing C-124 Globemaster."

Shortly after that, all my shared photos were missing from the Facebook page: *https://www.facebook.com/MissingC124/*, and the ones on Flickr were gone.

I wrote several letters to my Senator, Congressman/woman, and the President because I couldn't believe it was going to take up to six years to actually go and start recovering the deceased.

Dear Mrs. Castor

My name is Tonja Anderson and I am writing to you for help. On June 10, 2012 on Colony Glacier, about 40 miles east of Anchorage, Alaska a C-124 plane was found. Amongst this wreckage remains found. There were 52 airmen that lost their life for this country and there are family member still waiting for answers from our government on why they did not go back and retrieve these airmen remains almost 60 years ago. Mrs. Castor, I know that our military is going to say the crash site was inaccessible and ma'am, I did not buy this statement 12 years ago and I am not buying now. In 1952 the USAF landed at the wreckage to identify the plane and its remains therefore it was at some point accessible.

JPAC is the branch that is handling this find. However I do know they have other project ahead of the C-124 crash and will not get around to these remains anytime soon. Mrs. Castor, I know there has to be another DNA office that can do the testing that can move this process along. There are family member in their 70's, 80's and maybe there 90's who have waited all their lives to here that the USAF has found this plane crash and their loved ones. I am asking would you please aid in helping these families and me in getting results.

These men were Son, Husband, Fathers, Uncles nephews and grandfathers; that gave their lives for their country. Please don't let this country thank them again by putting them on a shelf until time is found to return them to their loved ones.

Sincerely

Tonja Anderson
Grand-daughter Airman Isaac Anderson Sr.

Dear Mr. President,

My name is Tonja Anderson and I am writing to you and the White House for help and inform you of the anger have about the military finding. On June 10, 2012 on Colony Glacier, about 40 miles east of Anchorage, Alaska a C-124 plane was found. Amongst this wreckage remains found. There were 52 airmen that lost their life for this country and there are family member still waiting for answers from our government on why they did not go back and retrieve these airmen remains almost 60 years ago. Mr. President, I know that our military is going to say the crash site was inaccessible and Sir, I did not buy this statement 12 years ago and I am not buying now. In 1952 the USAF landed at the wreckage to identify the plane and its remains therefore it was at some point accessible.

Since the finding of the plane I cannot believe that our Military would have no respect for the remaining families. These people have been waiting all their lives to hear we have found your brother, uncle, grandfather, son, or husband. For the military to state there is no video of what has taken place at the crash site or it was taken by a private company is bull. I as a family member should have NEVER had to go on the web to see all of the photo and video taken place out there for the reports to get before I had a chance.

The moment the airmen were brought off that glacier, put on land, given the proper military salute as a fallen soldier and placed on a plane headed to the lab for Identification; the families should have been able to see this. Once again this should have NOT been placed on the internet for the world to see, unless the families were told here is link to live feed of this ceremony.

At what point in time did the United States lose focus on what is right and what is wrong? The right thing to do was at least let the families know. They wrong this to do was put it on the web for the reporters, world, etc... to see. These we human beings and STILL have family members alive to whom they have had contact with.

JPAC is the branch that is handling this find. However I do know they have other project ahead of the C-124 crash and will not get around to these remains anytime soon. Mr. President, I know there has to be another DNA office that can do the testing that can move this process along. There are family member in their 70's, 80's and maybe there 90's who have waited all their lives to hear that the USAF has found this plane crash and their loved ones. I am asking would you please aid in helping these families and I in getting results.

These men were Son, Husband, Fathers, Uncles nephews and grandfathers; that gave their lives for their country. Please don't let this country thank them again by putting them on a shelf until time is found to return them to their loved ones.

Sir, I already know you will do NOTHING but also understand I will keep doing something until they right this wrong. You have lost my vote for this reelection and I will make sure the Airman Anderson's Son, Grandchildren, and great-grand children not cast their vote in your favor. While you preach on the campaign trail you have no clue on what is really going on and what really matters to the middle and lower class people.

Sincerely

Tonja Anderson

Grand-daughter Airman Isaac Anderson and a person at one point believed in you Sir.

On June 27, 2012, they publicly confirmed that the crash site found on June 10, 2012, was the C-124 Globemaster that went missing on November 22, 1952. I made a post on the Facebook page that was created in 2010 with the anticipation for the day that I would finally be able to present this news.

Families and Friends, what we all have been waiting for has come true. It would have been 60 years this November. There are no words I can state on this page or any other page that explains how I'm feeling right now. I would hope that each of you speak with the News to get our story out there. I would like to give a special thanks to the team that found the crash. Had they not went back we would not have closure right now. Based on what I'm seeing

in the photos the crash site was almost to the lake. Had it made it there we would not be coming together as one today. I pray and ask for JPAC to bring our Airmen home to us; one airman at a time or all at once just home to the families member waiting. ~Tonja Anderson-Dell

The Facebook page became the avenue for everyone to come together. We all shared our stories and became close friends. All of my letters, emails, and phone calls did not go in vain. One day they will all be home. This is what kept me motivated and moving forward. That I could help people I've never even met have closure.

I spent day and night reading posts from family members. Talking on the phone with ones not aware of any information of the crash other than they lost a husband, father, brother, grandfather, uncle, cousin, etc. I thought it was very interesting when some of the families stated they were told to contact me as I knew more about this crash then they did.

Then the photos started coming in one by one.

Michael Williams ▸ Missing C-124 November 22 1952
July 4, 2012

Howard Martin with his family

Earnest Dell, John N Patti Frigiliana and 3 others   1 Share

Lisa DeGiorgio Cover ▸ Missing C-124 November 22 1952
September 7, 2012 · Littleton, CO

We were just notified about my grandfather Edward Schnore today who was on this plane. Any information you can help me with would be appreciated. If you by any chance have a picture of my grandfather that would be great since my mom only has 1 picture of him.

Kevin Caid ▸ Missing C-124 November 22 1952
September 17, 2012

US Air Force S/Sgt Robert Dale Van Fossen, Greenbrier Arkansas

 **Kurt MacKenzie** ▸ Missing C-124 November 22 1952
June 26, 2012

 Meghan Coen ▸ Missing C-124 November 22 1952
October 28, 2012

I just found this page, what a great idea Tonja! My great uncle, Colonel Eugene Smith, was a passenger on the flight. My dad and Gene's brother have sent in the DNA kit back in August. Gene's brother Mike has shared many stories with my family about Gene's life in the Air Force. The local newspaper in Delaware (where Gene grew up) did an article regarding Gene and the plane crash in August. I have enjoyed reading everyone's posts and thank you for all the updates.

John N Patti Frigillana ▸ Missing C-124 November 22 1952
December 6, 2012

Linda Buie ▸ Missing C-124 November 22 1952
January 11, 2013 · Idaho Falls, ID

I just got word from Tonja that she is sending me a piece of the plane. I hadn't felt certain that I wanted it, but when I told her that, she said she would get me a piece and hold onto it until I did.

She is sending it today, and while my feelings are still ambivalent about it, I am very grateful that she was there to help me out.

Thank you so much, Tonja. You are a saint to be doing all of this for all of us. I used to go online and search for anything about that crash every so often. I was waiting for you and your website to come along. You have done a wonderful service to all the relatives of those who were lost.

I'm re-posting the wedding picture of my mom and dad taken on their wedding day in July of 52' just months before the crash. Tonja, you will never know how this website has helped me to connect to a father I never knew. With heartfelt thanks,
Linda Buie

 Linda Kittle Erickson ▸ Missing C-124 November 22 1952
September 4, 2012

PVT. Leonard A. Kittle

Cindy Cleveland Bell ▸ Missing C-124 November 22 1952
July 16, 2012 · Hamilton, MT

WESTERN UNION

AA129
A WA2 95 RX GOVT PD=FAX WASHINGTON DC 25 619P=
MRS GRACE WILES=
1913 EAST SCOTT ST PENSACOLA FLO=

THE SECRETARY OF THE ARMY HAS ASKED ME TO EXPRESS HIS DEEP REGRET THAT YOUR SON PVT BUIE, REGINALD IS MISSING IN ALASKA 22 NOV 52 PASSENGER ABOARD AIRCRAFT MISSING ENROUTE FROM MCCHORD AIR FORCE BASE WASHINGTON TO ELMENDORF AIR FORCE BASE ALASKA PERIOD CONFIRMING LETTER FOLLOWS=
    WM E BERGIN MAJOR GENERAL USA THE ADJUTANT GENERAL OF THE ARMY=

Brian Gorman ▶ Missing C-124 November 22 1952
November 10, 2012 · Wilmington, DE

This Photo was already posted by Meghan Coen, Gene's Great Nice. Uncle Gene was a lead commander of the CID (Criminal Investigation Division. He was the lead investigator of the Hesse Jewel robbery in Germany at the conclusion of WW II.

 **Joseph J Dodson** ▶ Missing C-124 November 22 1952
July 4, 2012

Jamie Ray Swift ▶ Missing C-124 November 22 1952
August 28, 2012

Here is my latest effort to add a picture.

 Terry K. Philips, Terri Adair and 5 others   2 Comments 11 Shares

 Stephen Seeboth ▶ Missing C-124 November 22 1952
June 30, 2012

My uncle was the Navy man on board.

> **Aircraft debris on glacier identified as 1952 wreck**
> Investigators say aircraft wreckage discovered this summer on a glacier in the mountains east of Anchorage came from an Air Force plane that crashed in 1952, killing everyone on board.
> ADN.COM

Missing C-124 November 22 1952 and Earnest Dell   2 Comments

👍 Like   💬 Comment   ↪ Share

 Glen Cleveland ▶ Missing C-124 November 22 1952
June 30, 2012

My brother was on that plane. My sister and I will never forget that week of the hope and hopelessness we went through while the plane was missing. It was a painful Thanksgiving Day and even more painful Christmas. We received the gifts he purchased for us while he was in San Francisco on his way to Anchorage during that week. It will be good to finally have closure.

6   5 Comments

Wim Koki ▶ Missing C-124 November 22 1952
June 28, 2012

http://www.foxnews.com/.../alaska-glacier-wreckage-is-150s-.../... my grandfather was on that plane

John N Patti Frigiliana   9 Comments

👍 Like   💬 Comment   ↪ Share

Belinda Mize Davis ▶ Missing C-124 November 22 1952
June 29, 2012

My Uncle, William Edmond (Eddie) Mize (19) was aboard this flight. If any further information is known, please contact me. We want to bring him home! Thank you!

Rebecca Wyatt   3 Comments

👍 Like   💬 Comment   ↪ Share

Simone Hickman ▶ Missing C-124 November 22 1952
June 28, 2012

My grandfather was on this flight and if anyone would like to contact my dad about this his name is Bob Hickman his father was Sterns, his email is roberthhickman@earthlink.net

Earnest Dell and John N Patti Frigiliana   1 Comment

Joseph J Dodson ▶ Missing C-124 November 22 1952
June 29, 2012

My cousin, Wayne Dean Jackson was a 21 year old steward on that flight. The recent discovery brings waves of relief to our family, and hopefully to all who had relatives.

5   1 Comment

👍 Like   💬 Comment   ↪ Share

Jamie Ray Swift ▶ Missing C-124 November 22 1952
June 29, 2012

This is Jamie Ray Swift again. I failed to mention that my father, James Herbert Ray, Jr., was one of 13 children and has brothers, sisters, nieces and nephews still living, some of whom I know are willing to donate DNA.

3   2 Comments

👍 Like   💬 Comment   ↪ Share

Simone Hickman ▸ Missing C-124 November 22 1952
June 28, 2012

Major Earl Stearns USMC with sons from left to right Greg, Bob, Earl Jr.

 Lisa DeGiorgio Cover ▸ Missing C-124 November 22 1952
September 9, 2013 · Littleton, CO

 Calvin Hall ▸ Missing C-124 November 22 1952
June 21, 2012

This is another angle on Colony Glacier where it enters Lake George, and likely includes the recovery location of the C124. This was taken on the flight up to the crash site in Sept. 2010.

Throughout my journey, I feel blessed to have talked to other family members of these wonderful men. I hope to someday meet them all. At the very least, I hope all of us can perhaps finally find closure together and celebrate our loved ones' sacrifice.

Chapter 6 - The Plane Has Been Found

Eventually, I decided that I wanted to book a flight to Anchorage, Alaska and see the wreckage for myself. I had everything planned for my trip and posted in the Facebook forum to let everyone know. I told them I would post daily while I'm there and include photos of my findings. If the weather permits I would be flying over Colony Glacier and Mt. Gannett.

The weeks leading up to the trip seemed like a lifetime. I even packed a week early. I was so excited to be in the exact some spot as my grandfather was. Even though I never knew him, somehow I felt that this would bring us closer together.

As I sat in the airport waiting to start my journey with my best friend, Janet.Strohsack, I was unsure of this expedition and what it had in store for me. As we boarded the plane and took our seats, so much was running through my head.

She asked, "Tonja, are sure you want to do this? Are you ready for this journey?"

All I could say is, "Yes! I have been waiting over 13 years for this day."

During the first leg of the trip, I dropped a tear. I wished my grandmother was still living to see this day with me.

As we got closer, I was able to see the mountaintops from the plane window, and my heart started pounding. Things were starting to feel surreal; I was less than 24 hours away from meeting the people who made my dreams come true. My dream of seeing what was recovered from the glacier, and getting into a plane that would take me on the same journey my grandfather took the night they all died.

facebook 👍

    Hello families...tomorrow is the big day. At the end of every night, I will post photos and updates. If you have any questions you would like for me to ask, please post them by tomorrow night because I will be on the base all day on Friday, meeting with the Alaska National Guard and the Joint Task Force.

    Stephen: Give them a salute and a thank you from the Seeboth family.

    Jamie: thank you and a 'bless you' will do. It will be an emotional journey (even for those of us not able to go) that we can share through your eyes, your heart, your photos, and your updates. Know we are all appreciative of any information we can get. Be safe.

    Mike: Make sure you stop by the airfield and say hi to the Black Hawk crews. Hangar 6

    Tonya: Give them a huge 'Thank You' from the Card family. I'm excited for you to go, you have been so wonderful to take time for not only me but for everyone. Your search has yielded such a huge reward. Have a safe trip; we will all be thinking of you, I have no doubt about that.

    Barbara: I'm excited for you and look forward to all your postings and pictures. I wish I could be with you in Alaska. Thanks for going. Barbara

    Vicki: Please give the Alaska National Guard a big thank you for Airman Jackson's family too. Good luck. My heart will be there with you.

    Michael: Hoping the sky is crystal blue, the wind is absolutely calm, and the weather perfect for your adventure. When I get far from home, a thought always passes through my mind "you're a long way from home." Have a safe and successful trip.

DAY 1

facebook 👍

Families, today was such a very emotional one. It took everything in me to hold back the tears. To walk into a room with members from most of the teams that helped us become one, was priceless. To hear the timeline from the moment they first sighted the crash site to the moment they brought our loved ones off the glacier was truly amazing. I stood there holding pieces from the plane, hearing what each piece was, and wondering to whom some of the stuff belonged to. I would later go to the section where they stored the wreckage debris. They opened container after container, and I stood there in disbelief. How could such a big plane

stand here before me in all these small pieces? I reached into each one of the containers and touch a piece, trying not to cry in front of the team. I have spent 13+ years looking for answers and trying to get a flag for my grandmother, but I have walked away today with so much more. The icing on the cake!

I want so much to thank Lt. Bryan for putting this together. I want to also thank each of the members who took time out of their morning to come and help me understand and to bring all of this to each of you.

~ more tomorrow, goodnight. Tonja

DAY 2
facebook 👍

    Yesterday, I was not able to take my trip to Mount Gannett and Colony Glacier due to snow and rain. The funny thing is, I was not ready to wake up to snow outside my window. Yes, I knew I was in Alaska, but I thought I made sure I did not go during their winter time. I was very sad, so we did a glacier tour and were able to see all the glaciers the military spoke about in the report. Throughout the report, they spoke of Surprise Glacier. I was able to see this glacier close up. "Wow," is all that I could say. I had a chance to see and hear a section break off, which sounds like thunder. I wonder if Colony Glacier is anything like this one

DAY 3
facebook 👍

    This morning, I woke up to fog and clouds in the sky. I had this heartbreaking feeling that I would not be able to see Colony Glacier, but I was wrong and I was given the moment of a lifetime. As I viewed this beautiful place, all I could think about was the 52 men on board the C-124 plane. When the plane came around and I was able to view Mt. Gannett, I just whimpered. I was flown to the crash site I closed my eyes for one moment. I said to myself, at this point in time, I'm as close to my grandfather as I will ever be. Airmen, you are not forgotten, and you are truly missed.

Tonja

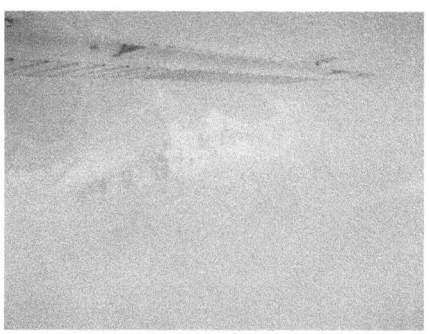 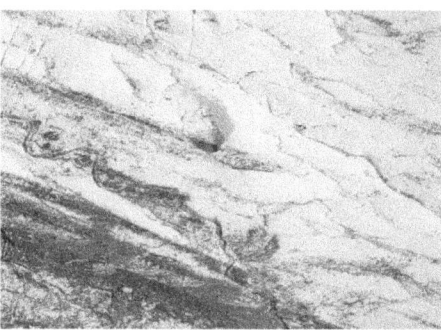

I wanted so much to be able to give the other family members the same experience I had while in Alaska. I took as many photos as I could to display the beautiful place that held our loved ones for so many years.

The pilot circled several times, allowing me to take as many photos as I wanted, and I was so grateful. When we circled for the last time, I looked down and saw something on the ground. We went around one more time and there on the glacier was an orange item. We both looked at each other and asked, "What do you think that is?" It was another part of a life raft.

We continued on our way back to Eagle River where my friend Janet was waiting. I had so much to take in on our drive back to Anchorage. I remember how the wind came off the mountains and the plane dropped a little in the air when we were flying above the glacier. I had so much running through my mind...will I die the same way as the others? Is this them showing me they know my journey and that they see me? Was it my grandfather showing his unconditional love? It was at this time I'd said to myself, *at this moment I'm the closest to my grandfather as I will ever be.*

Janet said to me, "Well, how was it?"

I told her, "Janet, oh my God, it was amazing, and I wish you'd been able to see it," I pulled out the camera to show her just how beautiful this place was. We made our way back to the hotel, and I made my post to the families.

Post to the families about my flight

I was given a view of the flight course as it made its way into the mountain. I looked around and all I could see was white snow everywhere. One peak looked like another. I have looked at photos of this site time after time and for years. At one point, I could not tell one peak from another. For this mount to hold someone dear to us all, it was a beautiful place.

I would then take the same journey to the crash site took as it moved down Colony Glacier for 60 years. Rounding the mountain step by step, I looked down on the glacier, looking for any sign of something new. Only because I knew from the meeting on Friday, they cleared the glacier of all debris. We came around for another attempt for me to view the glacier, and I saw something new. It looked like another one of the survival kits, still intact. I was that "kid in the candy store again"...laughing out loud.

After my post on Facebook, I added photo after photo for them to see everything as I'd seen it. I was hoping they could feel the same feelings I had when seeing the glacier. The comments posted by the families made feel like I had done my job.

Barbara: I get tears with every posting. Thank you so much being there and sharing.

Vicki: Thank you for sharing this with us. It is beautiful.

Tonya: Thanks for posting everything. You have written everything so beautifully. I tear up simply reading.

Jamie: Your words express your emotions so beautifully, I almost feel like I'm there...many of us ARE there with you in spirit and love. Thank you for "comforting" us... and God bless you all.

It was finally time to head home, but as we left our hotel room for the last time and made our way back to the airport, I couldn't help but think this is where my grandfather will forever be, until they find his remains and bring him home. So, as we boarded the plane and I took my seat, I say Goodbye, for now, Grandpa. I couldn't help but feel that I was leaving a piece of my heart in Alaska.

facebook 👍

Hello All, I have made it home. I will sit down to post more photos and provide details of the trip. I want to thank the USAF-JBER, Arctic's Air Academy, and Mike & Aaron Rocerata for all they did and tried to do for me to make this an amazing one. God Bless you all!!!!

I spent the next month catching up on calls and emails from other family members that have been contacted by the military for DNA. I found it exciting that some were told about the Facebook page and to go check it out. The stories told by them were powerful and brought tears to my eyes. What started out as a mission for a flag and answers became a journey and fight for not just me, but for everyone.

I asked God if this was his will and if he was sure I was the right person for the job.

Chapter 7: The Battle

It's every call, letter, and e-mail that makes this journey well worth it. Over the long fight for our fallen, families posted lots of photos and articles of their loved ones in the Facebook forum so we could all keep their memories alive. This brought us closer together, and we formed a small community to keep everyone up to date during the investigations. Most of all, we wanted to know if the ID process had started on the remains. In October 2012, the USAF had four families left to locate, and the Army had three. Meanwhile, we waited for the process to start, and I began reaching out to the military to see if it was possible to get pieces of the plane for the families who wanted them. For the families of the 52 men some of us may never welcome our soldier home and that piece of the plane maybe all we have.

My husband overheard me talking with another family who was providing their mailing info. He asked if the pieces would be coming to our home. I stated yes, and he replied "Oh hell no. I have watched *Ghost Whisperer* with you. Sorry, but I don't want 52 men walking around our home." All we could do was laugh.

Mr. Cronin from the Air Force Mortuary Affairs was able to step in, contact the right people, and worked with everyone to make this happen.

The first social media post came on Jan 2013 from Michael Williams. He and Ms. Fran received their piece of the plane. Then Mr. Gorman posted his pieces; Mrs. Deborah Jones posted hers; I received mine, and it took my breath away. Even after 60 years, I could still smell the diesel fuel on it. It was the calibration part with three posts under the lid. The more I saw the post, the more real this became. The huge plane that held thousands of pounds of cargo and 52 men had become all these small parts.

The families were so caught up in getting their piece of the plane we almost didn't see the press release from the Department of Defense:

IMMEDIATE RELEASE

Release No: NR-324-14
June 18, 2014

CORRECTION: DoD Announces Casualty Recovery

The Department of Defense announced today 17 service members have been recovered from a C-124 Globemaster aircraft that was lost on Nov. 22, 1952.

U.S. Army Lt. Col. Lawrence S. Singleton, Pvt. James Green, Jr., and Pvt. Leonard A. Kittle; U.S. Marine Corps Maj. Earl J. Stearns; U.S. Navy Cmdr. Albert J. Seeboth; U.S. Air Force Col. Noel E. Hoblit, Col. Eugene Smith, Capt. Robert W. Turnbull, 1st Lt. Donald Sheda, 1st Lt. William L. Turner, Tech. Sgt. Engolf W. Hagen, Staff Sgt. James H. Ray, Airman 1st Class Marion E. Hooton, Airman 2nd Class Carroll R. Dyer, Airman 2nd Class Thomas S. Lyons, Airman 2nd Class Thomas C. Thigpen, and Airman 3rd Class Howard E. Martin have been recovered and will be returned to their families for burial with full military honors.

On Nov. 22, 1952, a C-124 Globemaster aircraft crashed while en route to Elmendorf Air Force Base, Alaska, from McChord Air Force Base, Washington. There were 11 crewmen and 41 passengers on board. Adverse weather conditions precluded immediate recovery attempts. In late November and early December 1952, search parties were unable to locate and recover any of the service members.

On June 9, 2012, an Alaska National Guard (AKNG) UH-60 Blackhawk helicopter crew spotted aircraft wreckage and debris while conducting a training mission over the Colony Glacier, immediately west of Mount Gannett.

Three days later another AKNG team landed at the site to photograph the area and they found artifacts at the site that related to the wreckage of the C-124 Globemaster. Later that month, the Joint POW/MIA Accounting Command (JPAC) and Joint Task Force team conducted a recovery operation at the site and recommended it continued to be monitored for possible future recovery operations. In 2013, additional artifacts were visible and JPAC conducted further recovery operations.

DoD scientists from the Armed Forces DNA Identification Laboratory (AFDIL) used forensic tools and circumstantial evidence in the identification of 17 service members. The remaining personnel have yet to be recovered and the crash site will continued to be monitored for future possible recovery.

For more information, please contact the service public affairs office. Army public affairs office can be reached at 703-614-1742. Navy public affairs office can be reached at 703-697-5342. Marine Corps public affairs can be reached at 703-614-4309. Air Force public affairs can be reached at 703-695-0640.

Note: The correct rank for the Air Force service members is Airman 1st Class Marion E. Hooton, Airman 2nd Class Carroll R. Dyer, Airman 2nd Class Thomas S. Lyons, Airman 2nd Class Thomas C. Thigpen, and Airman 3rd Class Howard E. Martin. Additionally 703-614-1742 is the correct number for the Army public affairs office and 703-697-5342 is the correct number for the Navy public affairs office.

http://archive.defense.gov/Releases/Release.aspx?ReleaseID=16781

Seeing the press release made things so surreal for me. Because I knew I have not received a call from the Air Force that Airman Anderson was ID but still needed to read it over and over again to see if his name was there. What I did know was that when it came to locating and the ID of the recovered remains of the men, I would face a lot of difficulty.

What a lot of people didn't know was that in the last months of 2013 I was laid off from my job of 14 years and money was very tight. Now that the men had been found I wanted to make sure I attended every service I could. I sat down with my husband, and we found ways to make it work. I pulled out all of my money from my IRA, created a Gofundme page even though my pride didn't let me share it on social media a lot, did carwashes, and some families paid for my hotel room during my stay. If there was a will, there was going to be a way for me to attend.

While doing the planning, I wondered about the 2014 recovery, but I wanted to give JPAC time because I knew there was some changes to the agency. Changing the name to DPAA (Defense POW/MIA Accounting Agency), naming Army Lt. Gen Michael S. Linnington as director, and dealing with all the issues overshadowing the agency. After not hearing anything, I started researching and looking on the web when I came across a report called "U.S. Grave Science Marked" by Risk Aversion and "Bureaucracy" by Ms. McEvers with NPR. This story was done in March of 2014, and it provided a lot more detail into what was going on with the remains.

"MCEVERS: Then whatever's found at the site is brought back to JPAC. The non-biological stuff comes to this lab. Pieces of watches. Is that a lighter? That looks like some kind of jewel piece, piece of a bracelet, another buckle.

The biological stuff, the actual remains of the soldier go to the main lab downstairs. So this is the main JPAC lab. And just to describe it what we're looking at is basically just like a series of tables, one, two - what, 15 tables, yeah? And what's laying out on each table are remains, mostly bones. Some of them look to me, a layperson, like almost a complete skeleton. Others, this one that's right in front of us, it's fragments, pieces. In some places almost like dust. We walk up to one of the tables. Forensic anthropologist Marin Pilloud is getting to work on a case. So you're opening a box.

MARIN PILLOUD: Yes. This is a mummified foot. These were found in ice, so they're pretty well preserved.

MCEVERS: The case is a plane crash from the time of the Korean War. The remains of 19 servicemen were found in a glacier.

PILLOUD: Actually I have several guys in soft tissue just over there.

MCEVERS: This is one of the straightforward cases. The serial number on the plane led JPAC to a roster of who was on it. Still, Pilloud says it's taken her a year and a half just to sort the remains, test them for DNA and get those test results back.

PILLOUD: So right now they're separated by bones that I was able to refit back together. And then these tags all represent different DNA samples."
http://www.npr.org/templates/transcript/transcript.php?storyId=287328727

At this point, I reached out to the Joint PIW/MIA Accounting Committee (JPAC) in Hawaii and asked them about the remains. They directed me to contact Dover, where they claimed the remains were, but when I reached out, they claimed they didn't have them.

July 17, 2015

With all the changes over the past couple of months, I am not sure if you are the right person to contact. If you are not please forward this email to the proper person. It has been over a year now and as of today there has been no news on the remains collected last years. Where is DPAA in processing those remains and updating the families? Also moving forward who will be doing the recovery missions?

Thanks in advance Tonja Anderson-Dell

July 30, 2015

CLASSIFICATION: UNCLASSIFIED
CAVEAT: None

Aloha Ms Anderson-Dell,

Let me begin by apologizing for my response being so tardy. I have been out of the office for the past two weeks. As you have stated, there have been many changes over the past several months.

One of our teams conducted the recovery of this site this year; however, it appears this will be the last year we conduct the recovery operations. The determination has been made that the recovery of these individuals is outside of our stated mission since this was not a combat loss in a theater of operations. The Armed Forces Medical Examiner's Office has the responsibility for the recovery and identification of the remains from this site. The remains that we recovered last year along with the remains recovered this year have been turned over to the Medical Examiner for identification.

I hope I have answered your questions, if not please get back to me.

Respectfully,

Well you answered that best that you could. I reached out to them this morning and they have stated that NO remains from 2014 is in their possession. The only items they have are all items from 2015 recovery.

When did they get the 2014 remains?

Thanks for your help Tonja

No ANSWER!!!!

September 21, 2015

I wanted to reach out to you about the call I got on Friday evening from a reporter. Is it true that the remains were tested in 2014 and AFME is the reason we are in a standstill? Is there someone I can speak to at on your end that can address my question? Below is part of the story ran this weekend with a statement from Lt. Col. Holly Slaughter, the spokeswoman for your agency.

The Department of Defense POW/MIA Accounting Agency said it has known the identities of remains found at the glacier since at least March. But it hasn't released those names because it is no longer authorized to do so.

"The remains recovered in 2014 were sampled for DNA in July 2014, immediately after their arrival in our lab in Hawaii," said Lt. Col. Holly Slaughter, a spokeswoman with the agency, in an email to Alaska Dispatch News. "DNA results began to arrive back to our lab during the fall of 2014. By March 2015, final reports of new identifications were available, but due to the jurisdictional change, the identifications of the remains were deferred for the Armed Forces medical examiner to make."

Meanwhile, the identities of remains found on the glacier in 2014 haven't been released to their families. The Armed Forces Medical Examiner System would only say that it has received the remains.
Thanks in advance, Tonja Anderson-Dell

CLASSIFICATION: UNCLASSIFIED
CAVEAT: None

Hello Ms Anderson-Dell,

The article is accurate. Since there had been a change of jurisdiction, all the information we had obtained from the remains was passed to the Armed Forces Medical Examiner to officially establish the identification of the remains/individuals. I want to ensure I understand your question. Are you asking to speak with someone to determine why the jurisdiction was changed or are you asking why we did not proceed with the identifications. I'm not trying to put off answering you question, I just want to ensure we answer your question/s.

Respectfully,

September 22, 2015

Thanks for the reply. I think/feel I need an answer to both of those questions.
When the remains were found in 2012 JPAC's mission statement was the following:

Mission statement "Provide the fullest possible accounting for our missing personnel to their families and the nation"

Also

EXPLAIN HOW THE AGENCY CAN ASSIST IN A HUMANITARIAN CRISES? WHAT EXPERTISE CAN THE AGENCY OFFER IN THESE CIRCUMSTANCES?
The agency mission and the experience of staff members uniquely suit the agency to assist in many crises around the world. The agency has the largest forensics anthropological laboratory and the largest staff of forensic anthropologists and odontologists under one roof anywhere in the world -- several of whom hold the highest board certifications in their fields

Based on those statements there should have not been a transfer of agencies in my eyes. JPAC assisted the State of Alaska in a crises/ recovery. Someone should have stated this or fault harder to keep things states quo. I also find it hard to believe that AFMES never heard of our case until 2015. Is this TRUE?

Thanks in advance,
Tonja Anderson-Dell

CLASSIFICATION: UNCLASSIFIED
CAVEAT: None

Dear Tonja,

You make some very good points and I will try to address them. Our mission has not changed; however, there are some details that are not evident. I will do my best to explain in simple terms.

After the June 2012 discovery of the exposed wreckage, the appropriate leadership in Alaska was notified. The Joint Task Force-Alaska (subordinate to PACOM), being cognizant of our mission and capabilities asked PACOM if JPAC could respond to the situation.

As an organization under PACOM, JPAC undertook this mission in 2012 and returned after the spring melt again in 2013 and 2014. All three of these missions were conducted within the PACOM Commanders authority. Technically it did not fall into the realm of Personnel Accounting as our mission. It should have been Air Force Mortuary Affairs Operations (AFMAO). Typically AFMAO supports remains recovery at current operation crash sites, and does not excavate using archaeological standards. The matching of PACOM's authority with JPAC's capability drove the decision to use JPAC.

When JPAC merged with other entities to become DPAA, our teams were no longer under the authority of the PACOM Commander. DPAA's authority to conduct recoveries is limited by law to direct association with our past wars. This realization came to close to the dates available for this year's recovery. Once it was identified that shifting the mission to AFMAO at such a late date brought unacceptable risk to the mission, the Deputy Secretary of Defense ordered that DPAA once again conduct the mission. He also indicated that from this point forward (2016 and beyond) that the Air Force is responsible for the mission.

I hope this answers your question on the jurisdiction. Please free to call me if you have any questions.

Respectfully

September 23, 2015:

Thanks for the explanation. I have one more question. Once JPAC extract the DNA for testing, who does the testing? Is it done in house or sent out and then returned to JPAC? If sent out who does this for JPAC?
Thanks,
Tonja

September 24, 2015:

Hello Tonja,

We cut the samples in our lab and send them to the Armed Forces DNA Identification Laboratory at Dover Air Force in Delaware. Once they obtain the results of the testing they prepare the DNA report and send to us.

Respectfully,

Every government agency I talked to about the remains kept directing me somewhere else, and I kept getting the feeling that I was being lied to and ignored, so I took the next step I could think of and went to the media.

Missing C-124 November 22 1952
Published by Tonja Anderson-Dell [?] · July 26, 2015 · 🌐

Hello Families,

I am writing to you for help. As of June of this year it has been 1 year since remains were collected from the glacier by JPAC/DPAA. Those remains are still sitting and waiting to be ID. I have sent emails with no response. I am asking everyone to send an email to your congressman and senator asking for answers.

I was not going to go public to the families but fight this behind the scenes but I feel you all should know what is going on.

As of this morning I have wrote the Washington post and the President. I will write my congressman, senator, and my local newspaper tomorrow for help too.

The possible remains collected this year has been taken to Dover and will be processed there. They have nothing to do with the remains collected from last year and have no answers.

Thanks for all your help in getting our men home.

Tonja

Missing C-124 November 22 1952
Published by Tonja Anderson-Dell [?] · July 29, 2015 · 🌐

A copy my letter sent to the President and his staff. I have also sent letters for help to Ms. Kathy Castor and Mr. Jeb Bush.

Dear Mr. President and Staff,

My name is Tonja Anderson-Dell and I am the grand-daughter of Airman Isaac W. Anderson Sr; whom died in this crash on November 22, 1952. In 2012 during a training mission a blackhawk team found the crash site. In 2014 17 of the 52 men aboard the plane were identified and returned to their loved ones.

In June/July of 2014 JAPC returned to the glacier and remains were recovered. It is now over a year and these remains are still in the hands of JPAC/DPAA. I have sent emails requesting an update with no answer. In June of this year a team returned to the glacier and all items recovered there has been taken to Dover to start the process of identifying those possible remains recovered.

I have written to you and your staff in the past about JPAC now DPAA and nothing was done. Then later they got in trouble and light was shined on a lot of things they were doing wrong.

I am writing again in hope that someone will let you see this email and ask the proper people WHY is this lady writing me about this issue?
Most of all ask yourself if this was your loved one would you let this going without a fight.

Sir and staff, I mean no disrespect but enough is enough. How many more times will JPAC/DPAA mess up before you will put your hands in the pot?

If is seems like I am rambling it is because I am in the hospital and I still willing to fight for my cause.

Thanks in advance
Tonja Anderson

July 29, 2015, Between Sean with Alaska Dispatch and myself

Tonja,
It's Sean Doogan from Alaska Dispatch News -- I was wondering what is happening with the identification of remains recovered last year at the crash site.
I saw your Facebook post and wondered what was happening.
Thanks,
Sean

Sean,
Nothing has happen and that is why I have been posting. I feel the families should know the truth. I have sent emails asking for answers but no reply. Well you know me I will not go away that easy. If I have all the families ask for answers, I let the media know and they ask; just maybe they will do the job.
~Tonja

Tonja,
Whom do you suggest I contact at Dover regarding this? Please give me a call and we can get this done.
Thanks,
Sean

Sean,
What I do know...I contact JPAC?DPAA on July 17, 2015 asking for an update on the remains from 2014 only to be told they are in Dover and reach out to them. I contacted Dover via phone July 30, 2015 and was told that the remains are not there but still in HI. There is no reason to lie to me because they know I will research and find the truth. Even if that is putting them in the news...
I have sent emails to the President, Senator Bill Nelson, Senator McCaskill, several others asking for help but i know they will get the standard letter to send back to me.
Capt Edward Reedy (DPAA) and Lt Col Ladd Tremaine (Dover ME) are the 2 that should have this taken care of this. We (the families) don't have time for this as the family members are growing old and dying off knowing the plane has been found but not knowing if their Serviceman was identified. Since the finding of the crash in 2012 six (6) family members has pasted away.

All I am asking is if this were their serviceman wouldn't they want answers?
Thanks for anything you can do to help us.

~Tonja

September 21, 2015
Questions between Alaska Dispatch and Defense POW/MIA Accounting Agency (Public Affairs Rep)

Thank you so much for the help.
1. Here are my initial questions about the remains found on Colony Glacier, In Alaska, during the 2014 season.
2. (If true)Why did they sit at Hickham AFB in Hawaii for 16 months before being transferred to Delaware?
3. (If true) Why haven't the remains been tested yet (for DNA)?
4. What is the status of remains/artifacts found at the crash site from previous years and from 2015?
5. Have all the remains from those years been tested, and cataloged?
6. Were any remains or artifacts found this year? Have any family members been notified about remains/artifacts found since 2014?
7. Of the crew/passengers, how many have been identified through remains/or artifacts, and how many remain outstanding?
8. What parts of the plane were found this year (2015?)

Again, thanks for all your help.
–Sean

September 14, 2015
Questions between Alaska Dispatch and Armed Forces Medical Examiner System's Rep

Thank you so much for the help.

Here are my initial questions about the remains found on Colony Glacier, In Alaska, during the 2014 season.

(If true)Why did they sit at Hickham AFB in Hawaii for 16 months before being transferred to Delaware?

(If true) Why haven't the remains been tested yet (for DNA)?

What is the status of remains/artifacts found at the crash site from previous years and from 2015?

Have all the remains from those years been tested, and cataloged?

Were any remains or artifacts found this year? Have any family members been notified about remains/artifacts found since 2014?

Of the crew/passengers, how many have been identified through remains/or artifacts, and how many remain outstanding?

What parts of the plane were found this year (2015?)

Again, thanks for all your help.

September 21, 2015

Sean,
Here are our responses... let me know if you have any follow up questions. If you would like to use any of the following as quotes, you can use Colonel Ladd Tremaine, Director of The Armed Forces Medical Examiner.

Q. (If true)Why did they sit at Hickham AFB in Hawaii for 16 months before being transferred to Delaware?

A. The Armed Forces Medical Examiner (AFME) did not exert jurisdiction over the remains from the C124 Globemaster 1952 crash until the summer of 2015. The AFME was not made aware of the recovery and identification operation in regards to this incident until the spring of 2015.

Background: Jurisdiction by statutory authority outlined in title 10 USC 1471 establishes the AFME as the responsible medicolegal authority for the remains from this incident once the Alaskan medicolegal authority waves jurisdiction. The AFME received all remains from prior operations and the 2015 operation in August of 2015.

Q. (If true) Why haven't the remains been tested yet (for DNA)?
A. Samples from the recovered remains from 2014 did receive DNA testing. The AFME is currently cataloguing all remains and cross checking which remains were sampled.

Q. What is the status of remains/artifacts found at the crash site from previous years and from 2015?

A. All potential human remains and artifacts from prior recovery operations have been received by the AFME in August of 2015.

Q. Have all the remains from those years been tested, and cataloged?

A. The AFME is currently in the process of cataloging, matching remains to DNA reports, and determining if further testing is required.

Q. Were any remains or artifacts found this year (2015)?

A. Yes.

Q. Have any family members been notified about remains/artifacts found since 2014?

A. The AFME has not notified family members concerning recoveries made in the summer of 2015. The AFME was not a participant in the physical recovery of remains in the summer of 2015. The AFME is currently in the process of obtaining all the families demographic information from the respective service specific casualty offices and will actively update all concerned as to the process/status of identification.

Q. Of the crew/passengers, how many have been identified through remains/or artifacts, and how many remain outstanding?

A. The AFME was not involved in prior identification notifications and is currently in the process of obtaining all information concerning past identification notifications from the service specific casualty offices.

Q. What parts of the plane were found this year (2015?)

A. The AFME was not involved in the recovery operations occurring in 2015. The recovery operation, by policy, is the responsibility of the U.S.A.F., and the statutory authority of the remains is the responsibility of the AFME. Operations in 2016 will have AFME representation.

Again, let me know if there is anything else you may need.

When the media got involved, I started getting answers, but not in the way I wanted to. They were able to figure out that the remains of our family members were, in fact, identified, but there were two government agencies, JPAC and the Air Force Medical Examiner's Office fighting over jurisdiction of the remains. When they allowed the reporter to conduct a whole news report about the remains, they opened some of the body bags for her to see. This rubbed me and the rest of the families the wrong way. Someone from the media had access to our relative's remains before we were even notified that they had been identified.

LISA MURKOWSKI
ALASKA

COMMITTEES:
ENERGY AND NATURAL RESOURCES
APPROPRIATIONS
HEALTH, EDUCATION, LABOR AND PENSIONS
INDIAN AFFAIRS

United States Senate
WASHINGTON, DC 20510-0203

September 22, 2015

Honorable Christine E. Wormuth
Undersecretary of Defense for Policy
2000 Defense Pentagon
Washington, DC 20301-2000

Honorable Jonathan Woodson, M.D.
Assistant Secretary of Defense for Health Affairs
1200 Defense Pentagon
Washington DC 20301-1200

Dear Ms. Wormuth and Dr. Woodson:

I am writing today with great concern about an article that I read in the Alaska Dispatch News, our largest circulation newspaper, on September 18. The article concerns the identification and disposition of the remains of service members recently recovered from the November 22, 1952 crash of a C-124 Globemaster II in Alaska. There were 11 crewmen and 41 passengers on board at the time; none survived.

On June 9, 2012, an Alaska National Guard (AKNG) UH-60 Blackhawk helicopter crew spotted aircraft wreckage and debris while conducting a training mission over the Colony Glacier, immediately west of Mount Gannett, Alaska. Three days later another AKNG team landed at the site to photograph the area and they found artifacts at the site that related to the wreckage of the C-124 Globemaster. Later that month, the Joint POW/MIA Accounting Command (JPAC) conducted a recovery operation at the site and recommended it continued to be monitored for possible future recovery operations. The remains of 17 service members were initially identified and returned to the families for burial. However, subsequent visits to the Glacier have resulted in the recovery of additional remains which await identification.

The Alaska Dispatch article addresses remains recovered in 2014 that were initially turned over to the JPAC for processing. However, according to the article, it was subsequently determined that the JPAC was not the appropriate agency to undertake this work because the crash occurred on US soil. I am led to believe that the JPAC had to stop work and turn the effort over to Air Force Mortuary Affairs Operations and the Armed Forces Medical Examiners.

The article specifically indicates that the JPAC had already determined the identity of some of the remains but cannot release that information or the remains to families because jurisdiction over the matter now rests with another DoD component. It leaves the reader with the distinct impression that the Armed Forces Medical Examiners are reinventing the wheel, in effect identifying remains that were already identified by JPAC.

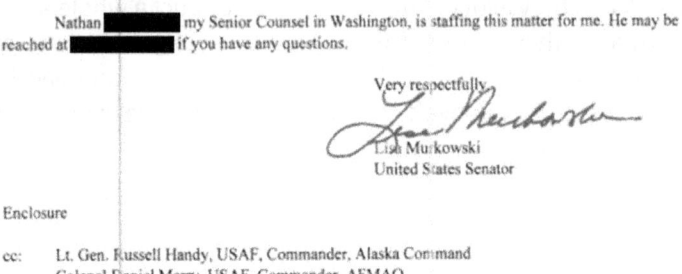

Even though the process of information was skewed, we were still able to get publicity about the crash, and it even caught the attention of the people at Capitol Hill. Senator Lisa Murkowski from Alaska reached out to me to figure out what was going on and decided that she would write letters to the two agencies to seek an answer as to why the families had not been notified.

Rep David Jolly called a meeting with several representatives to come down and hold a session at Capitol Hill. Since I was already so involved with the whole process, I decided that it would be my right to attend the meeting.

I arrived on Capitol Hill, and I got the chance to meet with David Jolly and other military personnel including U.S. Army Lt General Michael Linnington (DPAA), Army Col Ladd Tremmaine (AFMES), Boyd Sponaugle, Jennifer Vallee, Dr. Ed Mazuchowski, and other staffers. The meeting was called to discuss the issue with the remains and why I had to go to the media before I started getting answers. I asked why the families weren't notified about the remains found in 2014, and, since they were higher up in the Military, they were able to get me the answers.

They told me it had nothing to do with money, but more so with the confusion the two agencies had about jurisdiction. During the switch in command, the courtesy and requirement of telling the families that the remains were identified slipped through the cracks. So, because of an argument between two government agencies, the families of the fallen were left in the dark.

The whole reason this jurisdictional issue was occurring was because our men didn't die in a 'casualty of war,' and since we were not a war loss, JPAC couldn't legally have jurisdiction over the remains. So you are telling me because the men died in the United States, AFME believed that they should have jurisdictional rights to the remains. Our men were sitting out in the cold for more than 60 years and are now caught up in a petty fight between two government agencies that should've been dedicated to bringing those remains to the rightful families! Once this came to light, it was somewhat settled, and the government began to reveal who had been identified in the recovered remains of 2014.

I say somewhat settled because neither side admitted to the truth nor did they apologize to the families for the mistakes made. Since going to the media, things are moving along, and I/we am now kept in the loop about the decisions they make. They know that I will go back to the media if I suspect that they are withholding information from the families.

Currently, with the most recent mission in 2016, 42 of the 52 men have been identified. Because of the shifting nature of the glacier that the plane crashed into, the remains are constantly being shifted around. The entire crash site spans for miles, making recovery difficult. On top of that, the recovery men constantly risk their lives searching for remains.

Finding the remaining men can prove to be difficult, as there is a chance that they were shifted into Lake George. Because of the murky nature of the lake, the search team won't take the risk to find anything that might've gone into the lake.

I was once asked when I wanted them to stop looking for remains, as there is a chance that not everything will be recovered. To that, I responded that I didn't want them to stop until everyone was identified. There were only 10 men left to be found and identified, and my grandfather was one of them.

For the remains that are still up on that glacier, there's a chance that they will never be found or identified, and the families want some sort of closure for their dead relatives. What I feel the military should do is to take those unidentified remains and have a group burial for the families up in Arlington, which would be a huge honor.

Chapter 8 - Coming Home

June 22, 2014; PVT Leonard A. Kittle (Army) Caney, KS

As the door opened, I walked into a crowded room, and there stood Ms. Sandra and Ms. Linda. Ms. Sandra came running up to me, and we stood there just hugging and crying. Ms. Linda walks up and she says, "I thought you weren't coming." I told her "Thanks to Ms. Becky and Caney, KS Blue Star Mothers I was able to make it here to surprise you all." The look on their faces that evening made my trip well worth it. That night, I was able to speak to PVT Kittle high school friends and family. I arrived the next morning for his services, and when I turned the corner, the road was full of motorcycles. The local VFW, Patriot Guard, and the residence of Caney, KS lined the street as we made our way through. While making our way to the cemetery entrance,

there was a huge American Flag hanging for the motorcade to go under. What an amazing view!!!! To the Town of Caney, KS, Ms. Becky, Blue Star Mothers, Mr. Lawrence, the VFWs, and the Patriot Guards, I want to say "thank you" for everything you all did to Welcome PVT Leonard Kittle home. Ms. Sandra's son Earl made a comment to me, and I will forever remember it. Tonja, what we do in life echoes in eternity and what you have done for my mother and sister will forever echo into eternity. Thanks Earl ~Tonja

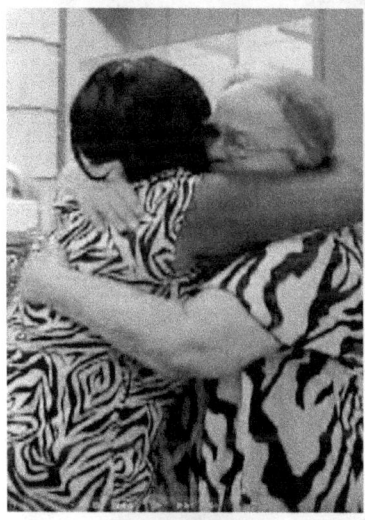

July 5, 2014, SSGT James H Ray Jr., Kittanning, PA

The weekend of the 4th of July I was able to attend the services of SSGT James Ray. The town of Kittanning, PA welcomed him home is a fashion to talk about for days. He was escorted by a helicopter, Fire truck, local and state officers. I was welcomed in like family and made to feel at home. The morning of the service I sat, watched, and cried because I saw his brothers and sister with tears of joy coming down their faces. The nieces, nephews, and cousins of SSGT Ray were there to see their parents experience a moment they had waited 62 years for. I stood in the cemetery taking in everything and saying my goodbyes to Jamie, Rubin, and Officer Denny when a man walked up to me and stated his name was Richard and he was the brother of Ray. He wanted to personally thank me for helping them have this moment that their parents died longing for. Mr. Richard, Mr. George, Mr. Donald, Mr. Perry, Mrs. Shirley, and Mrs. Martha: thank you for allowing me to share that moment with you. You're his witness now. Without a witness, they just disappear ... (Taking Chance) we are now their witness. ~Tonja

July 18, 2014, A/2C Thomas S. Lyons, Deerfield, FL

I arrived at the services of Airman Lyons, and it took my breath away. Interstate 95 was shut down for him. To look behind you and see cars for miles being held off by three police SUVs was amazing. His sister and children welcomed me with open arms. She and I got a chance to have some one-on-one time when realized we were holding up the services. I walked around viewing his life through pictures, and I have to say he was a great artist. We often wonder how such a tragedy like this could bring us all together so many years later. I think it held true for the Lyons's family. Lyons's cousin saw his story in the paper and read that he was coming home. They went through hell and high water to track down the family; from Alaska to Delaware. The two families have lived about 15 minutes from each other all these years. They attended the services, and it was a beautiful scene to see the cousins together like no years had passed by. The spotlight of the day was when a man walked up to Gerri, Lyons' Sister, and introduced himself. He was Lyons' friend from back in the day, saw the homecoming in the paper, and made his way to the services. He was even able to answer

some questions for them. There was a picture of a young lady in his wallet (that was returned after 62 years), and they wondered who she was. I want to thank the city of Deerfield, Broward Sheriff, Florida State Troopers and the Patriot Guard for welcoming Airman Thomas Lyons home where he belongs. The military pledges to leave no soldier behind. As a nation, let it be our pledge that when they return home walking or carried, we shall welcome them with open arms.~Tonja

July 19, 2014, Capt Robert W. Turnbull, Cario GA

After attending the services of Airman Thomas Lyons, we got into the car and drove 7 hours for me to attend the services of Robert Turnbull. I was due to speak at a gathering they held in his honor, but I was running late. However, I was still able to meet Capt. Turnbull's family and friends. In the room were items to show you the life of Turnbull as a son, brother, husband, and father. This included postcards and letters telling his late wife Ms. Doris how much he loved her and the children. One thing that caught my eye was an old green trunk. It was Turnbull's trunk from 1952 that he sent to the base prior to his arrival. What a remarkable treasure this was.
I spoke with his grandchildren and his daughter-in-law, Ms. Patsy because all his children had passed away. We drove down the street of a small town to bring Turnbull home. As we made our way through the street, cars pulled off to the side of the road to allow us to pass. This was emotional because they did this out of respect and were not asked by the police. We turned into the cemetery, and Jarrett (the grandson) had an American Flag hanging for Turnbull to pass under. Later that evening, I had a moment to spend time with the family; we all talked for hours about my 15-yearlong research, and I answered some questions they had. Thank you for welcoming my family into your home. Thank you to the town of Cairo, GA, Pine Park Baptist Church, Barnetts Creek Baptist Church Cemetery, Tyndall Air Force Base Honor Guard, Patriot Guards, VFWs, and the local sheriff department for welcoming Capt. Turnbull home to a family that truly missed him. Don't fight a battle if you don't gain anything by winning. I started the battle of bringing our men home, but I have won so much more.~Tonja

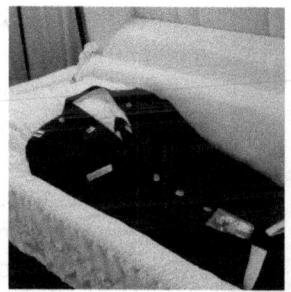

July 25, 2014, Col Eugene Smith, Wilmington DE

I was able to attend the services of COL Eugen Smith. For me, this was like no other ceremony I attended as he gave much more than just his life. I learned that COL Smith transferred to the newly formed Air Force OSI at Alaska Air Command. To see the members of the current OSI come out for his service was priceless. They brought COL Smith's remains into the church, he was escorted by one member from each one of his siblings. To see them stand tall knowing that the uncle they heard so many stories about was home and that they were escorting him in reminded me of a scene from a movie, where a hero comes home. They passed us and my son Tevin leaned over and remarked on how amazing this procession was. We stood in the cemetery, and I spoke with the family after the services. I am so glad that the younger generation kept some of them up to date on our page and my journey. We laughed and talked about social media and COL Smith coming home, but most of all we talked about the Generals, Colonels, and other officers that were there. I even got a chance to have some one-on-one time with one. What a wonderful person General Givens was. I asked if they came out because Smith was a Colonel and he replied that they came to pay their respects to the fallen soldiers, but also the man who helped start the OSI in which we serve under now. If your grandfather comes home, I hope that we will all be there to welcome him home too. We went back to the family home, and I got a chance to see more of the life of Smith. His personal items, old pictures, medals, his pin collection, and letters. Thank you for welcoming us into your home.~Tonja

September 20, 2014, 1Lt William T Turner (Crew-Navigator) Coudersport, PA

I met Lt. Turner's niece, Debra Jones, through Facebook and built a relationship with her. While attending the services of Airman Howard Martin, I was able to meet her in person for the first time. Linda asked if I would be able to make it to Turner's service and I told her if I could raise the money I would be there. And I did. His service was in Coudersport, PA, and the beautiful town was surrounded by mountains. The people there were ready to welcome Lt. Turner home. I was able to spend time with Ms. Mary, Lt. Turner's sister. I sat and watched her speak with distant family members and old schoolmates of Turner. I felt joy when I saw the smile on her face as she looked at her brother's casket. She turns to me and says "Tonja, he is home." The convoy of motorcycles, cars, fire trucks, and ambulances left the church, taking almost the same path Turner took as he walked to school every day. We made our way up the mountain, the locals from the different towns lined the streets and driveways waving flags and standing at attention. We turned into the cemetery, and the hearse drove him past the flag pole dedicated in his honor, and he was laid to rest next to his mother. "Years ago Lt. Turner's home became part of a National Forest, and you no longer have access to the road leading to his old home" I want to say thank you to Michael Wennin, Commissioner Paul Heimel, Mayor Brenda Whitman, General Frank Sullivan, and the Town of Coudersport for all you have done to welcome Lt. William Turner back home.

In Lt. Turner's yearbook, he mentions his long walks home and at his services the Pastor spoke on it as well. Lt. Turner, you were right, it was a long walk home. After 62 years I am honored to have taken that last long walk home with you ~Tonja

October 24, 2014, A/1C Marion E. Hooton, Sylacauga, AL

I was able to attend the services of A1C Marion Hap Hooton. At the cemetery, there was an American flag hung up in preparation for his service. No matter how many times I see this, it still takes my breath away. I was met by his cousin Ginger as she was in contact prior to the services. When word was sent that Hooton's caravan would be turning into the cemetery, everyone turned to see him make his way to his final resting place. His nieces exited the car and stood there with this look on their face that said everything that their voices couldn't. The bagpipes were playing in the background as honor guards moved him from the hearse to the gravesite. After the services, I got a chance to speak with one of Hooton's older cousins, Mr. Brooks. He thanked me for coming and just spoke about old times and the family's dedication to the military. I spoke with nieces and even got a chance to hear one of the many family stories about Hooton and about what a wonderful person he was. I only regret that I have but one life to give for my country ~ Capt Nathan Hale I want to thank you, A1C Marion Hooton, for giving that one life. ~ Tonja

December 17, 2014, 1 LT Donald A Sheda, Arlington, DC

The week of my birthday (December 20th) I was invited to attend the services of Lt. Donald Sheda by his daughter, Ms. Cathy. I arrived at the Old Chapel Church on Arlington's grounds, and the family was sitting in a room wondering if I would be able to make it. When I walked into the room, the look on Ms. Cathy face made my trip well worth it. I sat with her and her family before the services started and we talked about her journey of bringing her father home and how glad she was that I made it. I told her I wouldn't have missed it for the world. We were later escorted into the chapel of the church and prepared for Lt. Sheda's casket to come into the room. The doors opened, and the sunlight shone through; the guards lifted him and brought him into the chapel. I turned and watched Ms. Cathy stand proudly as her father was carried home.

After the services, he was taken outside and placed on the horse-drawn carriage. At that point, the family was informed that he will now be taken to his final resting place and that it would be a 2.2-mile walk. They also said that if anyone wanted to drive behind the carriage, they could. Ms. Cathy turned and stated, "No, I will walk with him."

My heart just quivered as I watched a daughter who had never met her father walk behind him so proudly and with such determination. Her family walked along beside her, making the trip through Arlington Cemetery. Visitors came to the edge of the pathway, took pictures and put their hands on their chest in a sign of showing respect. I turned to my son with tears in my eyes because it moved me so much. Arlington and their staff did an amazing job of welcoming Lt. Donald Sheda home. She had been proud of his decision to serve his country, her heart bursting with love and admiration the first time she saw him outfitted in his dress blues. ~ Nicholas Sparks

May 21, 2015, Col Noel E. Hoblit Arlington, DC

The son of a mason and boarding house entrepreneur joined the military (Reserves prior) in 1952, earned his Master in Dentistry, and was appointed the Chief Dental Surgeon at McChord AFB. Col Noel Hoblit has made his final journey home after 62 years, and I was there to see that journey. When I received the email from Hoblit's granddaughter, Heidi, asking if I would attend the services, all I could do was smile. Another one of our men has come home to his family after all these years. While I was checking into my hotel room, I noticed a face I had only seen in family pictures sitting in the lobby. I got to the counter, and Col Jerry Hoblit tapped me on the shoulder. He introduced himself and expressed his thanks. I will always hold on to the genuine words that he expressed to me.

The morning of the service was cold, wet, and dark. I made my way to the service and what I saw was a family that has longed for the return of their patriarch. The doors opened, everyone stood, and Col Noel Hoblit's procession made their way into the church. I turned, and the look on Col Jerry's and Fred Hoblit's faces was like two young boys waiting for their father to come home after a long day at work. After the services, everyone went to the gathering, and I was able to meet Mr. Fred Hoblit. He was only five years old when his father died. What I found so moving was when he stated that one of his good memories was of his father teaching him to comb his hair. I walked up to him and introduced myself. He had a big smile on as we talked and he reached into his pocket to show me something. Out came a little clear bag and inside was Col. Noel's dog tags from the plane crash. I had seen a lot of things from the crash, but this one was a little more personal. I asked if I could take a picture of it, if he didn't mind. He stated, "Yes, it is because of you I have it." I turned and said, "No, mother nature and the Blackhawk team played a big part too." We smiled and laughed. He and I talked a little more, and I asked if I could have a picture with him. As I am getting ready to leave, I took the time to walk up to Col Jerry to say goodbye, we talked a little more, and he introduced me to several other family members. I got a chance to hear some family stories, and we all took pictures. I want to personally thank the Hoblit Family for welcoming me with open arms and making me feel like part of the family. To Arlington, once again thank you for all you have done to return Col Hoblit to his final resting place. "One less day that I'll have to wait until I can take you in my arms and tell you again and again that I LOVE YOU." ~Col Noel Hoblit "Part of a love letter he wrote to his wife Ginia."

Col Noel Hoblit, I am glad I was there to see that one less day. He was laid to rest on the same day of his wedding anniversary to Ginia..

-Tonja Anderson-Dell

April 23, 2016, A/2C Bateman R. Burns West Helena, AR

The weekend of the 23rd I had the chance to attend the Celebration of Life Service for Airman 2nd Class Bateman R. Burns. Airman Burns was returned to his hometown in West Helena, AR. As I pulled up to the cemetery, I looked at all the people who showed up to welcome him after 64 years. Some

knew him, and some only knew the stories that were told about him. Those that didn't know him wanted to show their respects for the fallen soldier. One by one, family members came up and spoke about the man Airman Burns was: how he treated others growing up and every time he came home for leave. This   spoke volumes to me of the man   he was and the man he might   have been. I had the pleasure to   meet his youngest sister, Mrs.   Christine "Teenie" Manning. She   stood up, turned to the crowd   and told a story of their mother's   love. Holding out onto the hope   that he would one day return   home. Telling the family that   he might have amnesia but   one day he will remember and   return home. This is a story I have   heard so many times before.   My grandmother spent a big   part of her life thinking the   same thing. I sat and watched   as Nathan Burns (the eldest nephew/niece of Airman Burns) sat up in his chair, slowly bringing his hand up to his forehead while Airman Kelley presented the flag to him. I could not help but tear up even though I have seen this several times before.

A very proud nephew! I want to thank Airman Kelley for escorting Airman Burns home to his final resting place, Rubin with Air Force Mortuary Affairs for making this Celebration of Life complete, The Robert Darr Post 88 for allowing the family to gather after the Celebration, and Mrs. Vonda Burns for me. "You can easily judge the character of a man by how he treats those who can do nothing for him." ~ M.S. Forbes Airman Bateman R. Burns. I was able to hear your true character, and I am thankful to your family for allowing me to experience this.

Thank You for your Ultimate Sacrifice. Welcome home Sir

~ Tonja Anderson-Dell

May 25, 2016, A/2C Thomas J. Condon Waukesha, WI

I remembered when I received an email in 2011 from John Condon. He stated he was the nephew of Airman Thomas Condon and was hoping to get some information about the crash. He had been doing some research and came across the Facebook page. I was very happy to provide all the information I had on the crash and the crash being found as his father, William Condon, was coming to visit and he wanted to share it with him. When I heard the news that Airman Thomas Condon was identified and was coming home, it put a smile on my face. Another set of siblings lived long enough to see their brother make it home to his final resting place. On the 25th of May, Airman Condon made his way home to Waukesha, WI. Walking  into St. John's Catholic Church, I stood in line to greet the family, and I looked around at all the people that came to welcome him home. Row after row was filled with people. In the last row was the daughter and her family of PVT Kittle. Kittle died on the plane with Airman Condon and my grandfather. This was a special moment for me because PVT Kittle's homecoming was the 1st service I attended in 2014. I stepped out of line and gave Ms. Linda a hug. It meant so much for me to see them there. All these men died together aboard the C-124, and after their death, their families have become somewhat of a support system for each other. When it was my turn to finally meet his brother William and sister Marian, all we could do was smile at each other. Even if no other words were said, we were all glad to see each other. At the end of the service, a brother and sister walked behind their older brother's casket with smiles of joy. It is always nice to hear a little about the airman and their life before the crash. Airman Condon was a little funny because the story of how he loved to work on machinery was told, but once they did the math, they realized that he was only 10 or 11 years old when he found his passion. I want to thank Ruben Garza with Airforce Mortuary Affairs, Staff Sgt. Mario Super for escorting Airman Condon home, Fr. Bustos with St John Church, Tuscan Hall Banquet Center for allowing the family to gather and tell the wonderful family stories, John Condon for allowing me to attend and witness this all.

" Memorial Day this year is especially important as we are reminded almost daily of the great sacrifices that the men and women of the armed services make to defend our way of life." — Robin Hayes Airman. Thomas Condon, you have come home to your family close to Memorial Day, and it holds true to Robin Hayes Quote. Welcome home Sir and Thank You for making that sacrifice. ~Tonja

July 28, 2016, A/1C George M. Ingram (Crew-Load Master)
Beloit, Wisconsin

To the Ingram family, I am very sorry I was not able to attend the services. I did take the time to send flowers to let you all know I wish I were there in person.

"It is not light that we need, but fire; it is not the gentle shower, but thunder. We need the storm, the whirlwind, and the earthquake."
~Frederick Douglass

I will continue to be the fire, thunder and the storm until you all come home. ~Tonja Anderson-Dell

All photos were taken by Ms. Molly Beer Bussie

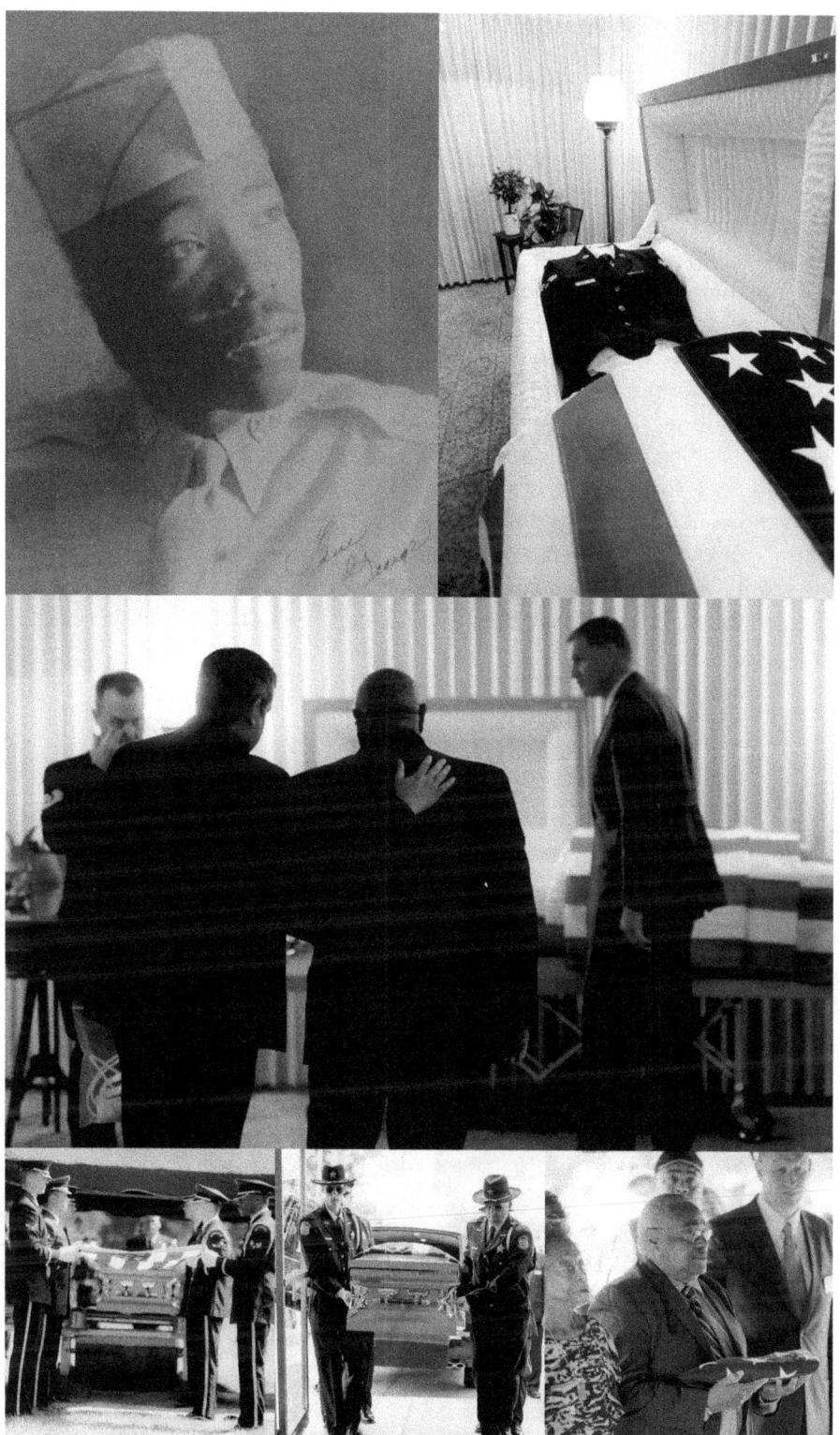

July 29, 2016, 2LT Robert E. Moon Arlington, DC

On the 29th of July, I went to the services of 2nd LT Robert Moon. It was a wonderful experience to sit and listen to his sister talk about his childhood and how he was her "knight in shining armor." The smile she had on her face when talking about it made me see just how much of a bond they all had. His cousin stood and sung a bidding farewell song to him with her lovely voice. As the family stood, turned, and walked down the aisle, I noticed a face I have seen before. I could not place it until later. I walked with them behind the horse-drawn carriage as they lay their knight in shining armor in his final resting place. It was very moving to see the familiar face stand and put his hand on his forehead. 2nd LT Robert Moon's cousin was Buzz Aldrin, and he attended the services to welcome his fallen cousin home. "Don't wait for the knight in shining armor. His armor is shiny because he has never been to war. Instead, look for the knight with torn and tattered armor, as he is the one who knows how to fight and is sure to be the one who can keep you safe from harm" ~unknown. To the family, Airman Moon's armor was torn and tattered. He spent his time showing you his love and his fight to keep you safe. I am so blessed to be allowed to see this Knight make his way home. ~Tonja

August 06, 2016, A3/C Wayne D. Jackson (Crew-Flight Attendant) Downing, MO

After attending Airman Moon's service, I spent the next weekend in Downing, MO for the services of Airman Wayne Jackson. Since I have been in contact with Airman Moon's sister, Ms. Vicki, this meeting would be a little more emotional.

I pulled up to the hotel with my son Tevin and my friend Janet. Ms. Vicki was waiting for us in the lobby, along with Airman Howard Martin's brother-in-law Michael. She and I hugged for several moments because it felt like this day would never come. We all sat in the lobby meeting each other's family before I had to go rest. It was a long trip for me, but her and I ended up talking till late in the night about our lives and how far this journey has come.

The service started out with a wake on Friday evening. People came out to show their respect to the young man that left the small town but returned their hero 64 years later. I got a chance to just sit and watch his classmates talk about him and meet his oldest living cousin, Ms. Ella B. She warmed my heart when she tapped me on the shoulder to tell me who she was.

The next day we lined up with the family and walked into the church; for a moment, I felt a little overwhelmed because I longed for the day I could bring my grandfather home for such a long time. I pulled myself together and gave Ms. Vicki the support she needed when she gave her speech. She spoke about what a wonderful person Wayne was and how they grew up together in his family home, even though they were not technically related. Her face showed just how much she loved and adored him when he came home on leave. He would take her for rides in his car, just doing things big brothers do.

We made our way through the town to take him to his final resting place. People stood along the way holding flags and waving to the cars as they passed. The local restaurant put a sign out welcoming Airman Wayne Jackson home. We watched as the honor guards pulled Wayne from the hearse and placed him between his parents in the family grave. Wayne finally made it home, two days before his mother's birthday.

A hero is someone who has given his or her life to something bigger than oneself ~Joseph Campbell

Wayne, you did just that, and I thank you for it. ~Tonja

Finding Closure - Synopsis:

I've accomplished a lot more than what I set out to do. There is still more for me to do. As you already know my journey started as a quest to get an American flag for my grandmother. I was able to get it shortly after her passing when the Mac Dill Air Force Base gave me the flag and even provided a funeral service for my grandfather. As of today, May 2017, 42 of the soldiers have returned home to their families leaving 10 Soldiers remaining. I will continue to fight to get them home, this includes Airman Anderson. I want to bring my grandfather home.

When I started my journey a lady (Ms. Wood) contacted me. We talked for a long time. Something she said has stayed with me. She stated, "I have lived years knowing my husband was one of the reasons the plane crashed." Her feelings of guilt troubled me. Yes, I have read the report. When reviewing the report, somethings made me feel that the pilot was not to blame. There were a lot of factors at hand that could have caused the plane to go down. Leaving the families believing that it was a navigational error just does not seem right or fair to me. I would love to be able to have the accident report corrected for the soldier's families. Col. Jack Stovall said it perfectly. "How can you say navigational and pilot error when you know that the radios weren't that good in that area and that the de-icing of the wings weren't good, the forecasting wasn't done right so you can't say it was navigational and pilot error."

My next effort, will be to see the report corrected to state it was not navigational error. If I can get the military to say that it wasn't the flight crew's fault and it was caused by a whole bunch of other factors, the families will get some closure. Most importantly, Ms. Wood will not have to carry that guilt anymore. She will know it wasn't her husband's fault. Even though Patricia, the pilot's wife, may have passed away, it's still important to me to have this corrected for her. After more than a decade, my journey still hasn't ended.

For all of the people who have reached out to me asking how did I do it? I can tell you this; do the research, document the journey, don't give up, never take no for an answer, and always remember to write your local Congressman, Senator, and even the President. These people ask for our votes, and they make us promises, It is appropriate that we ask them to make good on some of those promises. I love to research and discover the truth. If you every need my help, please feel free to contact me, and I will do my best to help.

Some of my biggest accomplishments during my quest were naming the peak the C-124 Globemaster crashed into after our soldiers. It is now forever named "Globemaster Peak."

UNITED STATES BOARD ON GEOGRAPHIC NAMES

In reply please use this address:
U. S. Geological Survey
523 National Center
Reston, Virginia 20192-0523

May 12, 2014

Ms. Tonja Anderson-Dell

Dear Ms. Anderson-Dell:

We are pleased to inform you that the U.S. Board on Geographic Names, at its April 30, 2014 meeting, approved your proposal to apply the new name Globemaster Peak to a summit on the ridge southwest of Mount Gannett in the Valdez-Cordova Census Area of Alaska. The name has been entered into the Geographic Names Information System, the nation's official geographic names repository, which is available and searchable online at http://geonames.usgs.gov. The new entry reads as follows:

Globemaster Peak: summit; elevation 8,983 feet; located approximately 55 mi. E of Anchorage, on land administered by the Bureau of Land Management; the name honors the victims of a recently-recovered C-124 Globemaster plane which crashed on the summit in 1952; Valdez-Cordova Census Area, Alaska; Sec 16, T13N, R7E, Seward Meridian; 61°12'53"N, 148°12'09"W; USGS map – Anchorage A-4 NE 1:24,000.

Sincerely yours,

Lou Yost
Executive Secretary
U.S. Board on Geographic Names

Letter from the President of the United States

THE WHITE HOUSE
WASHINGTON

January 8, 2017

Ms. Tonja Anderson-Dell
Tampa, Florida

Dear Tonja:

As Commander in Chief, I am profoundly grateful for your devotion to honoring the legacies of our fallen service members—proud men and women who safeguarded our freedom and risked everything to defend our country.

Selfless heroes like your grandfather endure in the ideals they advanced and in the values we hold dear. Guided by the pursuit of peace, these patriots helped move us toward a more hopeful tomorrow, and we can never forget the price they paid so people around the world might know justice, equality, and opportunity.

May God bless all who served our country and their loved ones. And may we carry forward the work of those who gave their last full measure of devotion and keep their memories burning bright.

Sincerely,

2012-2013 Recovery on Colony Glacier
Taken by a member of the United States Air Force

My trip to Alaska June 2016

2016 Recovery on Colony Glacier
Taken by Allen Cronin with AFMAO

Photos and Letters sent to me over the years about our soldiers

TSgt Hagen is sitting on the maintenance stand on the left looking back. The guy looking up on the right facing the camera is SSgt John J. Buckley an engineer and a great guy to fly with. The two guys looking up are A/3c Thomas Ford and A/3c Marlon Scott, both radio operators.
-provided by Chuck Schuster September of 2013

The pictures showing the barracks compares how things looked in 1952 when I lived in the third one from the right to how that same area looked in Nov 1996. It is now a parking lot. If you look to the left of both pictures you can still see the same permanent brick structure which is the HQ for the wing.

This letter was sent to me by the local newspaper. I wanted to share this as it gave me an insight as to what our men were doing hours prior to taking off.

Thank you Mr. Darryl

```
TO: Tonja Anderson-Dell         RE: C-124 Crash
    Tampa Fla.
       To Kieth ████ @ tampatrib.Com
       Tampa Tribune

Dear Tonja Anderson-Dell
   First I would like to introduce myself.
I am Darryl
                      (Age 85)
   I recently read your email written in 2012,
about the discovery of the C-124 Globemaster that
crashed in Alaska in 1952. Your story really hit home
with me, as I've been waiting 62 years to know the
story that you related in your email.
   Your devotion to your Grandfather, Isaac, is such a
meaningful story that I can relate to completely.
   I was stationed at McChord Field in Tacoma, Wa.
```

at the time of the crash and was a Loadmaster on C-124's hauling passengers and cargo to virtually all over the world.

The lost plane was from my squadron and I had made several trips on that same plane with most of the crew & Aircraft Commander. He was, without a doubt, the finest pilot I've ever met. We often flew together as he usually requested me as part of his crew, but I was away on a trip to the Arctic Circle and had just returned. I had 1350 hours flying on C-124's.

I can personally give you a first-hand accounting of the information released to us crew members at McChord. I was personal buddies with all of the crew and to many of the passengers at the time of the crash. I can relate more details if you request them.

Pg 2

I would like a complete list of the passengers and crew if possible which will help me remember better. I remember one friend who's name was Sgt. Ray, I went to Loadmaster School with him in Palm Beach, Fla.

The plane was being fueled & made ready to go to Anchorage, Alaska. I was having breakfast at the Mess Hall in Takoma at McChord when I spotted some crew members & a few buddies that I had gone to school with in Florida, waiting to depart.

They invited me to join them, so I did, and we spent about an hour visiting and laughing before they left.

The next morning I heard that their plane had crashed in Alaska. I didn't know any of the details at that time, but I was anxious to see if there were any survivors.

It was several days before we got a report, after the weather had cleared. We were told that there were no survivors.

That is the last I ever heard of this crash until May, 2014 when I read your email. I was absolutely amazed at what you found & reported.

Congratulations on your endless efforts to immortalize your grandfather, and the support you've given the other families.

Your grandfather was about the same age as me when the plane went down. He must have been a very special guy to have a granddaughter such as you my dear.

I'm very anxious to hear the latest on the past two years explorations at the Crash Site.
I wish you continued success in your search and I'll be sure to add your name to my prayer list

Pg. 3

along with the many others you have been there for."
Please remember the list of passengers & crew for me if you can get it.
If I can be of any other help to you, please contact me. You can also call me at Area Code (███) ███-████ or email me at / ██████@yahoo.com

        Your concerned friend,
        Darryl ████

(Formerly: Airman 1st Class Darryl ████
(1952-53) 1703rd A.T.G.
        Continental Div.
        MATS
        Tacoma, Wa.

Now! ████████

Let me know of any new findings, please

Letter from Steve Scott Sr., the brother of A3/C Marlon L. Scott

March 14, 2014

To: Mrs. Tonya Anderson[Bell]

Hi Tonya!

I am writing to you knowing that we have never met, but believe we are comrades in spirit. You should know that I originally came across your name early on in researching data for my book, "The Longest Flight Home",

At the time of the tragic crash of the C-124 Globemaster in the wilds of Alaska, I was finishing basic training and on to tech training in Communication Center Operations; then on to Korea. So, many intimate details escaped me for decades. It was at the time I was writing the book that I read an account that detailed your dogged pursuit on the USAF to get on with recovery efforts at the crash site. It is my understanding that your Grand Father was lost in the crash and your Grandmother was understandably, suffering from unresolved grief that spurred you to direct action. Your unremitting efforts in that regard are to be lauded by all families who lost loved ones.

Moving ahead years to the time of my publishing of the "Longest Flight Home", that unsuspected contact has been made and a web of information ensured. Several weeks ago, I received a phone call from a Mrs. Vicky Dodson. She had seen the book on my Face book page. To be honest with you, I am not much of a devotee of social media, but my "geeky", savvy grandchildren, gave me the face book as a Christmas gift.

It was at this point that Vicky gave me the names of three additional people she has remained in contact with and suggested that I do the same. There names are: Charles Shuster, Michal Williams and yourself. I have since contacted each of them and found all to be wonderful, dedicated individuals. Each has supplied me with intimate details that had escaped me through the years. It is a pity that I didn't know each of you at the time of publishing my book.

Again, I felt compelled to write to you and wish you well in your continued pursuit of finding a degree of closure; albeit it seems we all will have to await future outcomes. I am enclosing a copy of my first book, "A Walk Through Time." I hope you enjoy reading it.

Best Regards for a healthy

New Year!

Steve Scott Sr.

Loving memory,

JBER Historian Mr. Doug Beckstead
Sandra Kittle (wife of PVT Kittle)
Martha Ray Rider Ellenberger (sister to SSGT James Ray)
George Ray (brother to SSGT James Ray)
Ray Martin (brother of Howard Martin)
Isaac Owens (A/2C Robert Owens)
Veda Ponikvar (sister to Capt John Ponikvar)

SOLDIERS STILL MISSING

Anderson, A/B Isaac W. Sr.
Budahn, A/2C Verne Chester
Buie, 2LT Reginald "Reggie" (Army)
Coombes, CAPT. William
Costley, SSGT Eugene R. (CREW - 2nd E)
Draskey, CAPT Delbert D.
Goebel, CAPT Jerome H.
Kimball, A/3C James R. (CREW - FA)
Leaford, 2LT Jack R. Jr.
McMann, A/2C Dan F.
Miller, A./2C Edward J.
Mize, A/2C Edmond W. "Eddie"
Newsome, A/1C Sterling E.
Schnore, MSGT Edward J. (Army)
Scott, A./3C Marlon L. (CREW - Radio Operator)

ABOUT THE AUTHOR

Tonja Anderson-Dell was born and raised in Tampa, FL. She has a very close knit family and is a dedicated and loving wife, mother to five beautiful children, and grandmother to two wonderful grandsons.

While discussing her family lineage with her paternal grandmother she learned of her grandfather's death which sparked a desire to learn more. She has a BS as a paralegal and decided to put the research skills learn to work by researching about her grandfather and the other soldiers aboard the missing plane. The quest to learn more took her to places she never though she would go, Mt Gannett, Alaska. Taking the same path the men took the day they went missing. Through her research, she made some powerful friends and enemies. This book is a compilation of her years of research.

www.ingramcontent.com/pod-product-compliance
Lightning Source LLC
Chambersburg PA
CBHW071219160426
43196CB00012B/2348